LEARNING RESOURCES®

Three Bear Family™

Activity Book

Grades K-3

by Carol A. Thornton
& Judith K. Wells

Printed in the United States of America
Learning Resources
Lincolnshire, Illinois 60069

⚠ WARNING:
CHOKING HAZARD - Small parts.
Not for children under 3 years.

Table of Contents

Introduction

The *Three Bear Family™ Activity Book* provides problem-solving activities and ideas for the K-3 level using Learning Resources' Three Bear Family Counters (LER 725) or the reproducible manipulatives in the Teaching Aids section.

The learning and language experiences in this manipulative-based approach are designed to create a solid basis for mathematical understanding. Children learn to think and communicate mathematically as they manipulate and describe what they observe and do with the bears. The active learning situations presented in this book stress:

- exploration
- analysis
- representation
- communication
- prediction

As a supplement to the on-going, early school mathematics program, the activities enhance the development of a variety of concepts and skills, including:

- Free Exploration
- Sorting and Classifying
- Patterning
- Counting and Number Sense
- Estimating and Measuring
- Adding and Subtracting
- Multiplying and Dividing
- Fractions
- Problem-Solving
- Graphing and Probability

Think • Tell • Share

Throughout the *Three Bear Family Activity Book*, cooperative learning is emphasized. The following teaching-learning suggestion will be helpful:

Encourage children to:
- THINK about their response.
- TELL or SHOW their partner what they have thought or done.
- SHARE their findings or conclusions with the larger group.

To increase reflective thinking and discussion, first ask children to think of their personal response to a task. Give them time to tell their partner and act out a solution with the bears. Then, invite the children to share their findings with the rest of the class.

Using the Activities

Teaching Notes for each pupil page are provided at the beginning of each unit. They are organized into Warm-Up, Page and More For You sections designed to help you introduce, complete and extend the activities on the blackline masters. Class instructions and story problems that can be read aloud are printed in bold.

Teaching Notes
Free Exploration

Three Bears

Warm-Up: Start by asking children how the story of the Three Bears begins. Distribute the bear counters and allow time for the children to make their own discoveries. Then, ask questions that focus on the differences and similarities in size, shape and color.

Language Experience: Put several bears in a box lid. Have a child pick a bear and use attribute words to describe its color and size (ie. yellow, Baby Bear). Repeat the activity so all color and size words are reviewed.

The Page: Working with a partner, have one child tell a story while their partner uses bears to act it out.

More For You: Guess My Bear. Taking turns with a partner, one child hides a bear while the other asks up to 3 questions to try to guess which bear it is (ie. "Is it blue? Is it middle-sized?"). Students should keep a running tally of the number of questions needed to guess the bear. Low tally wins.

Know Your Bears

Warm-Up: Line up a Papa, Mama and Baby Bear overhead tile (or transparency model from Teaching Aid 1) on the overhead projector. Without removing them, place an additional bear tile on the projector and quickly turn off the light. Ask the children to think, and then tell their partner which bear was seen.

The Page: Using Teaching Aid 1, have children color, cut and paste any 9 bears onto the page's game mat. As a bear counter is drawn from a box and described, children place a marker on the matching bear if it is on their game mat. The first player to get three bears in a row (down, across or diagonally) wins. After a whole-group introduction, the game can be played by smaller groups or partners.

More For You: Family of Bears. Collect 2 bears of each size in 2 colors and put them in a box. Working with a partner, students take turns selecting a bear. The goal is to get a family of bears (Papa, Mama, Baby) all the same color. A player can keep no more than 3 bears at a time, so a bear will need to be discarded each time after the 4th draw.

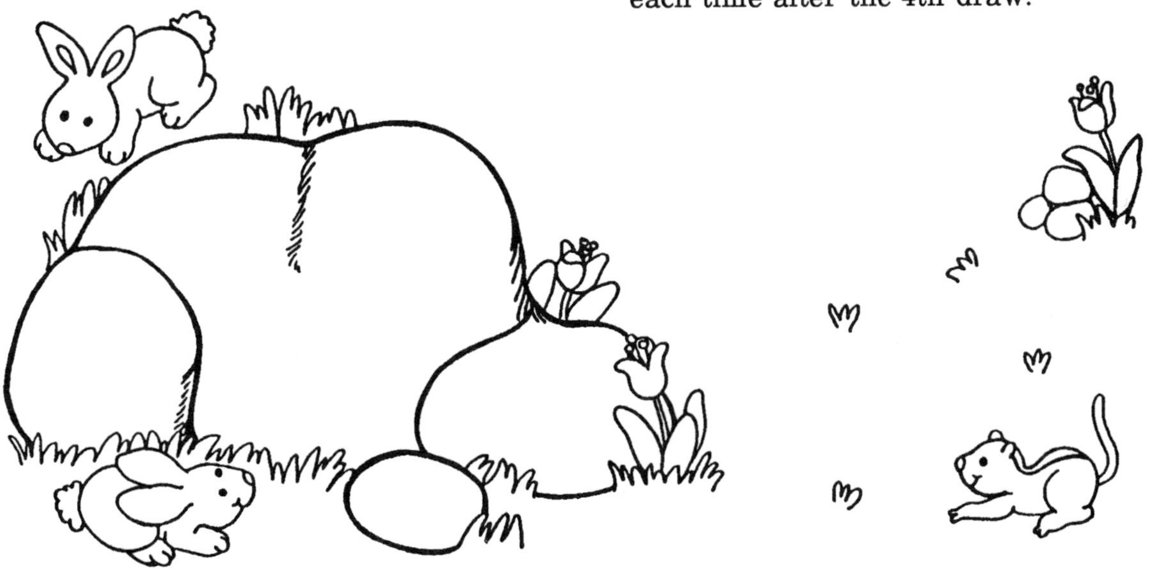

Three Bears

NAME ____

Tell a bear story.
Use bears to act out.

▶ **Next, play "Guess My Bear!"**

Know Your Bears

NAME ______________________________

Color, cut, and paste a bear in each box below.
As bears are drawn, cover the box if you have a match.
Try to get 3 in a row.

▶ **Next, play "Family of Bears."**

Teaching Notes
Sorting and Classifying

Bear Sort

Warm-Up: Give some bears to each small group of children and direct them to find as many ways as they can to sort them into two groups; into three groups; into four groups. Be sure children include "not" sets (ie. red and not red), as well as colors and sizes. Children should be aware that some groups may not be able to sort into four groups.

Overhead Sort: Place a handful of overhead bears on the overhead projector and sort them into piles. Select another bear, and ask children to tell their partner where the bear belongs. Call on one pair to share. Continue until all the bears are sorted.

The Page: Taking turns with a partner, one child thinks of a way to sort the bears, then puts each group in a different place on the mat. (It is alright if some parts of the mat are not used.) The other child guesses the sorting rule. Have children write the sorting rule. ***Variation:*** Working independently, children sort, write the rule, and remove the bears. They then exchange rules with a partner and re-sort the bears.

More For You: Is It Alike? Children work in pairs with an assortment of bears. Children begin by placing one bear, the GAME BEAR (e.g., a red Papa Bear), between them. At each turn and without looking, a bear is taken from the pile. If it is like the GAME BEAR in some way, the child keeps the bear. If not, the bear is discarded. When all bears are used, the child with the most bears wins. Children can then combine winning bears and tell/write/report something about them (e.g., they are all red bears or Papa Bears.)

Bear Hunt

Warm-Up: Distribute bears to pairs of children. Then say, "I went on a bear hunt and I found a bear that was NOT green. Can you find one?" (Elicit several responses.) Repeat the bear hunt with other "not" statements until children are ready to continue the activity with a partner.

The Page: For each row, children place appropriate bears over those shown. Have them think how the bears are alike and choose another one for the box that is different. Children compare and discuss with a partner and then with the class. Encourage different solutions. Students record by drawing, coloring, pasting (Teaching Aid 1) or writing to describe their choice.

More For You: Distribute bears to each pair of children. Taking turns, each child selects a bear and tells one thing that IS true about the bear. Then the other child tells one thing that IS NOT true.

Out You Go!

Warm-Up: Make finger puppets using bear outlines from Teaching Aid 1. Have four students show 3 blue Baby puppets and one red Baby puppet. In pairs, the rest of the class can tell each other which bear does not belong and why. Have them share their response with the class. Repeat, using some examples where there is more than one possible response. For example, show 1 red Baby Bear, 2 blue Baby Bears, 1 blue Mama Bear. (Possible answers: Red does not belong because all others are blue; OR Mama does not belong because all others are Baby Bears.)

The Page: For each row, children place appropriate bears over those shown and choose another bear so that ONE in the row does not belong. Children compare and discuss with a partner; then with the larger group. Encourage different solutions. Students record by drawing, coloring, pasting or writing to describe their choice.

More For You: Pick Four. Each child randomly picks 4 bears and tells a partner if one is different from the others. Children keep a tally of the number of times when one was different. The winner is the child with the most points when time is up.

Compare the Bears

Warm-Up: Hold up two bear puppets or show two overhead bears. Ask children how they are alike and how they are different. Repeat with other pairs until children are ready to continue the activity with a partner. Let children work for a while, then challenge them to find and show their partner a pair that is different in only one way.

The Page: Provide an assortment of cut-out bears and Three Bear Family™ Counters. Children color the bears on the page and beside each, place a bear counter that is different in only one way. To record, children color and paste a cut-out bear to match the bear counter. In the space provided, children write to tell how the two bears are different.

More For You: Pick a Pair. Taking turns with a partner, one child selects a pair of bears and tells how they are different. If they are different in only ONE way, player keeps bears and partner gets one point. If not, the bears are returned and no point is scored. Each player has 5 draws before the other takes over. The winner is the one with the most points when time is called.

Bear Parade

Warm-Up: Select one child to be the first in a parade of children holding puppets or bear counters in the front of the room. Have children think about what type of bear could come next and tell their partner. Have them share with the group and select a second child to join the one-difference parade. Continue until no bears that are different in one way remain.

The Page: Children work in pairs. One puts a bear in the star as the "leader" of a one-difference parade. Taking turns, children line up other bears on the footprints to form a parade where each is different in only one way from the bear next to it. Children may wish to continue the parade beyond the edge of the page. Have them color and paste bears to record.

More For You: Bear Ring. Start with a circle of four bears, each different in only one way from the one next to it. Players try to insert another bear into the circle, keeping the bears different in only one way. Children can score one point each time they can do so. The player that earns the most points, wins!

Bear Sort

NAME ______________________________

Think of a way to sort your bears.
Put each group in a different place on the mat.
Ask your partner to guess your sorting rule.

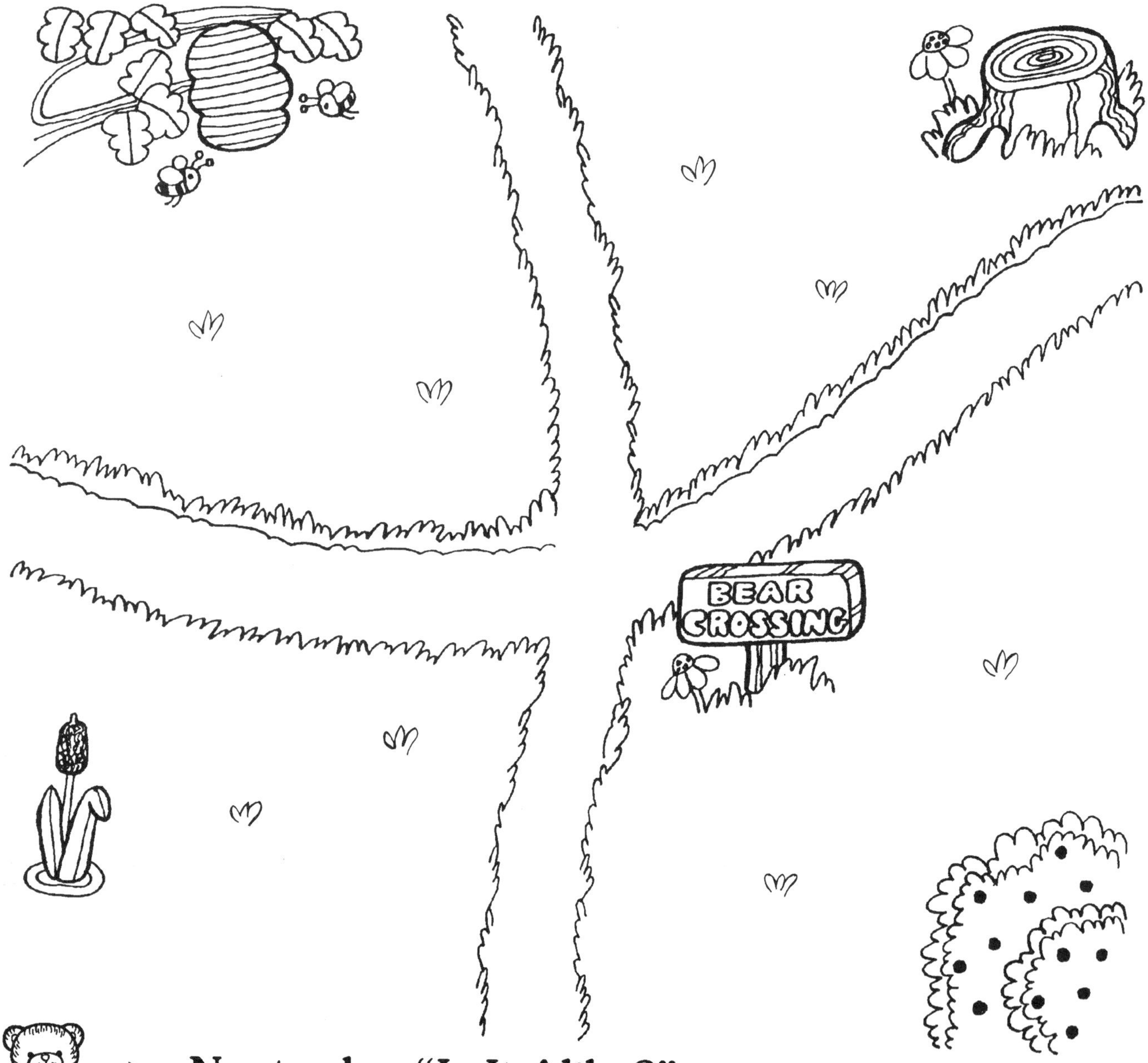

▶ **Next, play "Is It Alike?"**

Bear Hunt

NAME ______________________________

Think of how the bears are alike. Pick a bear for the box that is NOT like the others.

1.

red green green

2.

red red red

3.

blue blue blue

► **Next, play "Is — Is Not."**

Out You Go!

NAME ______________________________

Pick a bear for the box so that ONE in the row does not belong.
Mark the one that does not belong.

1.

yellow yellow yellow

2.

green blue green

3.

red blue red

▶ **Next, play "Pick Four."**

Compare the Bears

NAME ______________________________

Color the three bears.
Beside each, put a bear counter that is different in only one way.
Color and paste a bear to match.
Write and tell how they are different.

1. 

____

2. 

____

3. 

____

▶ **Next, play "Pick a Pair."**

Bear Parade

NAME ____________________

Take turns with a partner.
Choose a bear to lead the parade. Put it in the star.
Line up other bears for the parade so that each is different in only one way from the bear next to it.

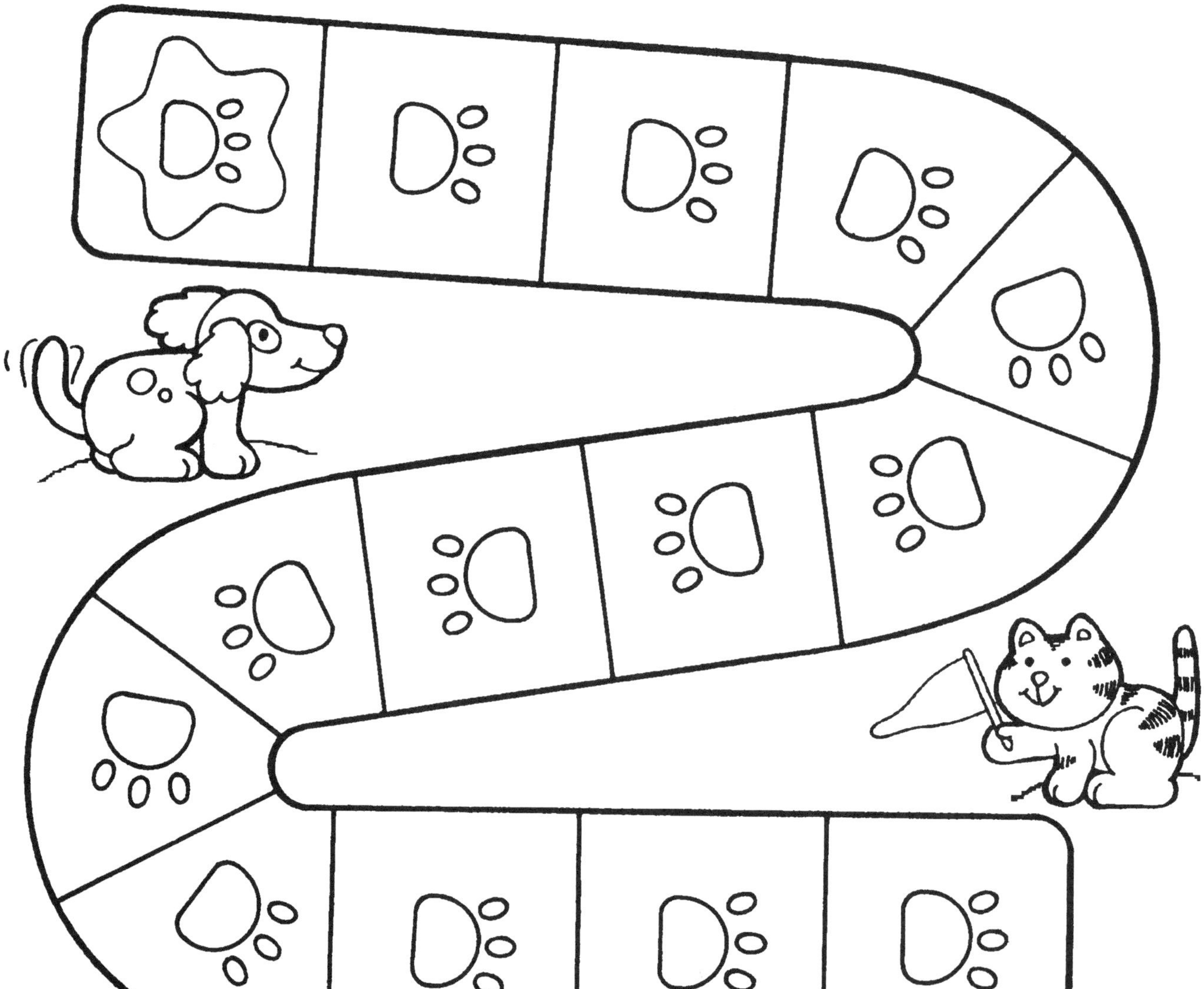

▶ **Next, play "Bear Ring."**

Teaching Notes
Patterning

Bear Patterns

Warm-Up: Distribute some bears. Begin a movement pattern (ie. clap, snap, snap; clap, snap, snap) and invite children to join in. Have children describe the pattern. Encourage them to consider different ways of expressing the pattern using letters, colors, sizes, positions, etc. Have children use bears to show the pattern. Repeat for different patterns.

The Page: Let students work independently to create a pattern, extending it off the page onto their desks. Then, have them describe the pattern in one or two ways to their partner, and listen as their partner does the same. Children decide and circle the correct response to show whether their patterns are alike or different.

More for You: All the Ways. Taking turns, Player A uses 2 or 3 bears to establish the core pattern. Player B uses bears to show different ways to continue the pattern. One point is earned for each way shown and described.

Start The Parade

Warm-Up: Work with the children to arrange a parade pattern of bear counters. Ask volunteers to describe the pattern and tell what would come next. Have children not in the parade close their eyes while the first three people in the parade place their counters behind their backs. Children then open their eyes and tell what is missing at the start of the parade.

The Page: Children first work independently to create a parade pattern. They then remove the three bears leading the parade. Children look at a partner's parade and try to identify the missing bears at the start of the parade. (Encourage children to vary their patterns by color, size, and position).

More For You: Take Two. Taking turns with a partner, one child makes a parade pattern and then hides ANY 2 bears. The other identifies the pattern and tells which bears are missing. Children score 1 point for each bear correctly identified. The winner is the player with most points when time is called.

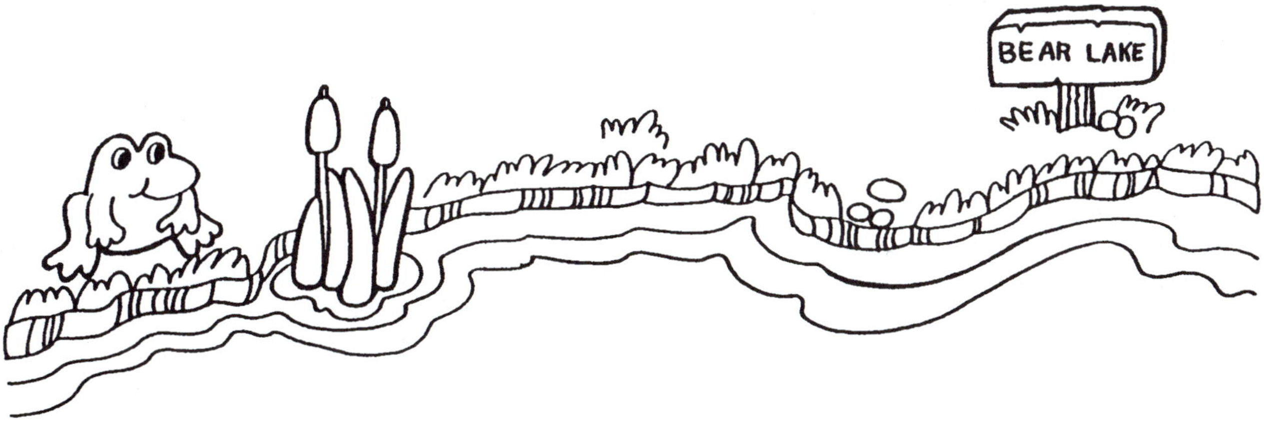

Bear Roll Call

Warm-Up: To introduce and review the terms *row, column, diagonal, corners, vertical, horizontal, down*, and *across*, provide directions which require children to move bears to different locations in the 3 by 3 matrix on the page. Model it on the overhead projector or alternately, draw a 3 by 3 matrix on the chalkboard and attach the bears with masking tape. For example:

"Place a green bear in the four corner boxes. Place a red Baby Bear in the 2nd row center. Place a red Papa Bear in the 1st row, 2nd column. Place blue bears in the 2nd row 1st and 3rd columns. Place a red Mama Bear in the 3rd row, 2nd column."

After all the boxes have been filled, have children tell their partner one thing they notice about the arrangement of the bears. In this case, all bears in the 2nd column are red. The color arrangement in the top and bottom rows are the same. Repeat this exercise with other arrangements. Children can work in pairs as they become more comfortable with the terms.

The Page: Working in pairs, children use bears to act out the story on the matrix:

"It was the first school day after the Great Bear Picnic. When Teacher Bear rang the bell, the 9 Baby Bears lined up in a yellow, green, green pattern on the bearprints. Teacher Bear said, 'Take your seats,' and all the bears sat down to talk about the Picnic."

Get students to look at the matrix and describe what Teacher Bear would see. Are there any patterns? (All in column 1 are yellow; all in columns 2 and 3 are green; each row is yellow, green, green.) Repeat to form other arrangements.

More For You: Write About It. Children look at the matrix they have colored and write about the patterns they see.

Seat the Bears

Warm-Up: On the overhead, place overhead bears* in a 3 by 3 matrix to form a pattern (e.g., Row 1: blue Papa, blue Baby, blue Baby; Row 2: yellow Papa, yellow Baby, yellow Baby; Row 3: green Papa, green Baby, green Baby.) Get children to describe the patterns they see: A-B-B; Papa, Baby, Baby; same color in each row and so on. Children can work in pairs as they become more comfortable with the terms. One child fills a matrix and gives clues to the partner so that the second child can, without looking, fill a matrix the same way.

* You can also make your own from Teaching Aid 1.

The Page: Working in pairs, direct children to use bears to act out the following situation:

"The bears went to see a movie. They were seated in a special way, as you can see. Some of the bears left. Find the missing bears and seat them."

Have children start by covering the pictures with bear counters. Children place bear counters in empty spaces so that a pattern is formed. Then they cut, paste, and color to record. (Accept all logical solutions).

More For You: I Take, You Tell. Taking turns with a partner, one child makes a matrix pattern and then hides 2 bears. The other identifies the pattern and tells which bears are missing. Children score 1 point for each bear correctly identified. The winner is the player with the most points when time is called.

Bear Patterns

NAME ______________________________

> Use bears to make the pattern.
> Describe it to a partner.
> Compare the pattern to your partner's pattern.
> Circle the correct word to tell if they are alike or different.

1.

alike   different

2.

alike   different

3.

alike   different

▶ **Next, play "All the Ways."**

Start the Parade

Warm-Up

NAME ____

Make a parade pattern.
Take away the first three bears.
Have a partner tell what is missing.

▶ **Next, play "Take Two."**

More for You

Bear Roll Call

NAME ______________________________

Listen to the story.
Put bears on the paw print to act it out.
Color to record.

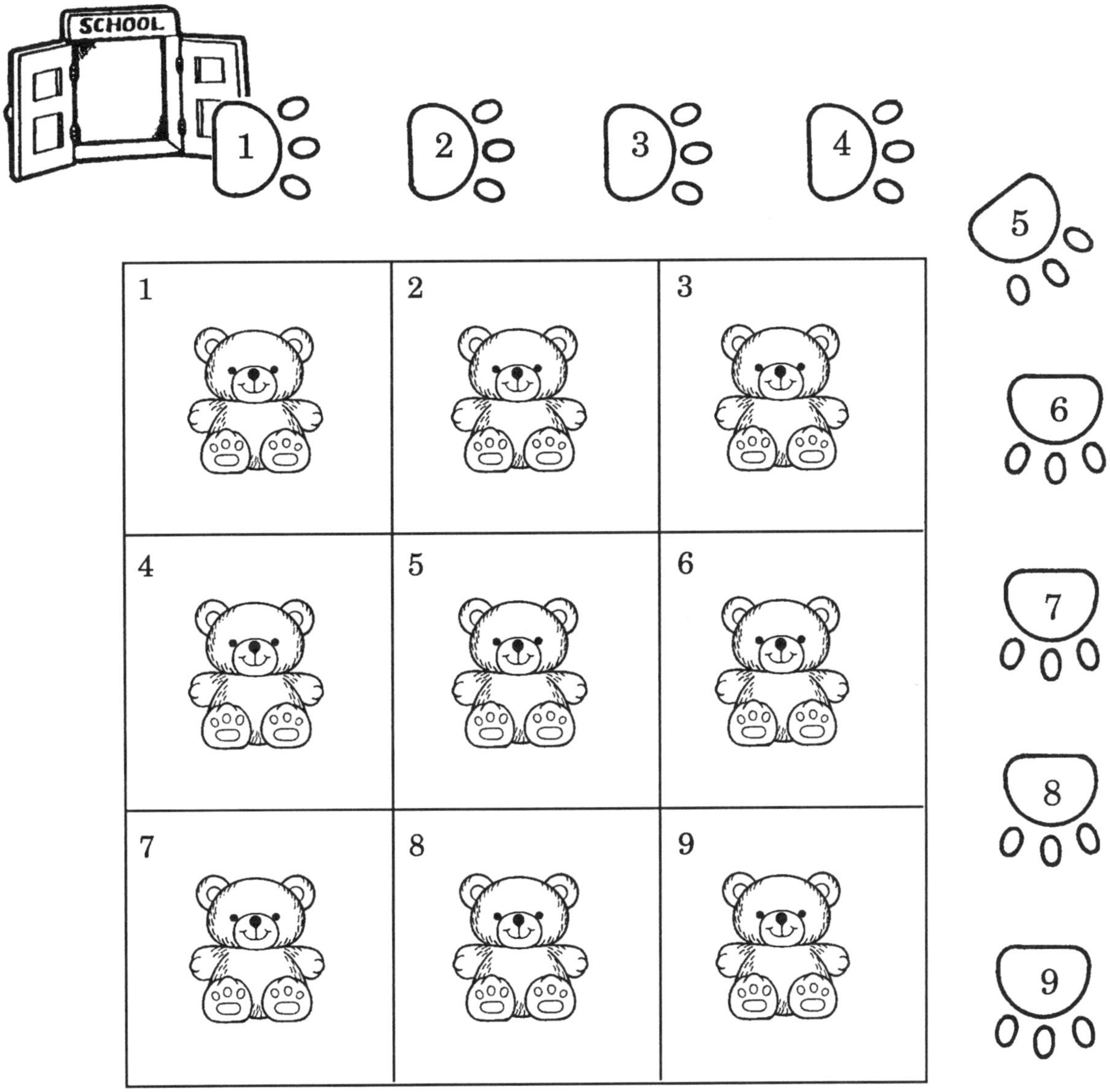

▶ **Next, do "Write About It."**

Seat the Bears

NAME ______________________________

Listen to the story.
Use the bears to act it out.
Cut, color and paste to record. Write about it.

Write how you know where to seat bears.

____

____

____

▶ **Next, play "I Take, You Tell."**

Teaching Notes
Counting and Number Sense

Bear Count
(Counting, Counting on, Counting back)

Warm-Up:
Counting — Distribute several bears to each child. Have children each pick up one bear. Say: **"Stand up if you picked a Mama Bear."** Have the class count the number of children standing and tell/write the number. Repeat for Papa Bear, Baby Bear and different colors.

Counting on — Call 6 children to come to the front of the room. Have 4 children stoop low. Direct the other 2 students to stoop as you tap them on the shoulder. **"How many are stooping?** (4) **Keep counting. Now there's 5, 6 stooping."** Repeat with different numbers of children OR count back as children return to their seats.

Counting back — Ask: **"How many children are at the front of the room?** (6) **Count back as I send them away** (tap them on their shoulder) **— so I know how many are still here."** (6 here, 5 here, 4, 3) Do not let the count go all the way back to 1 or 0. Stop the count earlier to emphasize counting back 1, 2 or 3 numbers. Repeat for other numbers.

The Page: Distribute several bears to each pair of children. Taking turns moving the bears, children listen to the story and follow directions. Tell the story without stopping so children can enjoy it. Retell the story stopping after each part to have children share and check the counting. Encourage students to match bears 1 to 1 to check whether their count is the same as their partner's. Select numbers and vary situations to fit children's needs.

Counting: **"Eight bears were having a picnic by the lake. They all sat down on the blanket. After they enjoyed the sun a while, three bears unpacked the picnic basket. The others had great fun racing from rock to rock. When Papa Bear came, he saw one bear on each rock."**

Continue with other stories. You may wish to ask for 1 or 2 more than what is counted; 1 or 2 less; or for a number that is a little more or a lot less than the number that is counted.

Counting on: **"Bear Troop #6 planned a water day at the lake. As soon as all the bears were there, three jumped into the water. Some more bears went in, too. The lifeguard kept track: "3 — 4, 5, 6, bears in the water."**

Repeat for other numbers and other lifeguards. Each time, count on 2 or 3 more.

Counting back: **"The Honey Bear Club had lunch down by the lake. 7 bears sat on the blanket to eat. One by one as they finished, some left. Mama watched and counted to herself: 7 — 6, 5, 4."**

Repeat for other numbers and different situations. Each time, count back 1, 2 or 3 more.

More For You: Will It Fill? Provide some bears and a Counting Frame (Teaching Aid 3). Taking turns with a partner, children fill their hand(s) with bears and tell whether they think there is enough to fill the frame. Have them count as they place bears in the frame to check.

Too Many, Not Enough, or Just Right

(More/Less than 5; More/Less than 10)

Warm-Up: Place a transparency of a 5-Frame* (Teaching Aid 3) on the overhead and beside it, a small group of bears. Ask children to estimate and tell a partner whether they think the number of bears is too many, not enough or just right to fill the frame. Count and use the frame to check. Each time, the number of bears should be recorded on the chalkboard by the child or teacher. Call on a child to verbalize the comparison of the number to 5 (e.g., 6 is more than 5; 3 is less than 5.) Repeat for other small handfuls of bears.

The Page: Distribute several bears and a 5-Frame* to each pair of children. Taking turns with a partner, ask one child to take a handful of bears while the other child counts and tells where to write the number. Children use the 5-frame to check.

More For You: Bear Hunt. Provide a copy of Teaching Aids 4 and 5* and some bears. To begin, ask children to place different color bears on START. Taking turns, have players draw a card and describe the number (e.g., 7 is 2 more than 5; 4 is one less than 5). If the number is more than 5 they move forward that number of spaces; if less, they move back. If exactly 5, no movement is made. Children may use the frame as needed to help or to check thinking. One point is earned each time the player passes START. The winner is the player with the most points when time is called.

* Or, when working on more/less than 10, use the 10-Frame from Teaching Aid 3. Follow directions parallel to those above. For the *More For You* extension activity, provide Teaching Aids 4 and 6.

Compare and Race to the Bell

Warm-Up: Call two volunteers to the front of the room. Have them each take some overhead bear tiles and count as they place them on the overhead projector and write the number of bears. Line up each group of bears 1 to 1 and compare. Emphasize language by asking children to verbally describe the comparison: "I have 7 bears, 7 is more than 5." OR "I have 4 bears; 4 is less than 8." OR "The group with extras has more. I have extra bears so I have more."

The Page: Distribute cards from either Teaching Aid 5 or Teaching Aid 6 and two bear counters (different colors, for game markers) to each pair of children. Ask children to draw a card and write the number shown. Have the children work together to compare the numbers and circle the one that is greater. The child with the greater number advances 1 space. The winner is the first to reach the bell. Children may use bears to help or to check.

More For You: Go for Less. Provide another copy of the page, cards from Teaching Aid 5 or 6 and two bear counters for markers. Children play as above, but the child with the number that is less, advances 1 space.

Papa Says, Mama Says
(Introduction of 2-digit numbers)

Warm-Up: Give children ample opportunities to count and count on by 10s. You may wish to have children count beads on a frame; count unifix 10-trains as they are put away, and so on. If your children have not had many counting experiences like these, try to provide them before proceeding with this next page.

Choosing the Counting Method: Provide a bag of 64 "berries" (round counters) for each group of children and relate this sample story:

"Papa Bear and Mama Bear sat down to count the berries they picked. Here are the "berries" they picked. Count and see how many berries."

Observe how children count to obtain a total. Do they touch and move the berries so as not to count one twice or skip a berry in the count? Do they count by 1s? by 5s? by 10s? Discuss different counting strategies. Challenge children by asking: **"How could we arrange these berries to make them easier to count?"** Guide children to discover that grouping by tens is an especially quick way of counting large numbers of objects.

The Page: Distribute 50-100 round counters for berries to each pair of children. Introduce the page with this story:

"On the first day of berry season Papa Bear brought in a bucket of berries from the woods and said, 'Take 2 big handfuls of berries and tell me how many you took.'"

In turn, have children take handfuls of berries, counting and recording the number. Children should be encouraged to work together to count. If children are counting by 1s, challenge these children by asking: **"How could we arrange the berries to make them easier to count?"**

Ask children to continue with the rest of the page, taking the number of handfuls specified.

More For You: More Berries. Working together, have children examine the berry count for each day and decide which amount is greater; or how many were taken each day in all. Listen and discuss children's responses.

Deliver the Mail

Warm-Up: Place a transparency of a Hundred Chart (Teaching Aid 7) on the overhead and use a peep-hole card to identify 30 as the starting number. Challenge children to count on or back by 1s; 10s; to tell the number before or after; to tell other numbers hidden by the card. Repeat the activities with different numbers.

The Page: Distribute copies of Teaching Aid 7. Introduce the page with a story like the following:

"The bear mail carrier has many letters to deliver, but some of the numbers are missing from the mailboxes. Help the mail carrier by writing in the missing numbers."

Children use the peephole cards to help find or to check the numbers they write on the "mailboxes."

More For You: Peep and Tell. Distribute copies of Teaching Aid 7. Taking turns, children place a peep-hole card over the number. The partner earns 1 point for every missing number that can be told. The winner is the child with the most points when time is up.

Bear Count

NAME ____________________

Listen to the story.
Use bears to act it out.

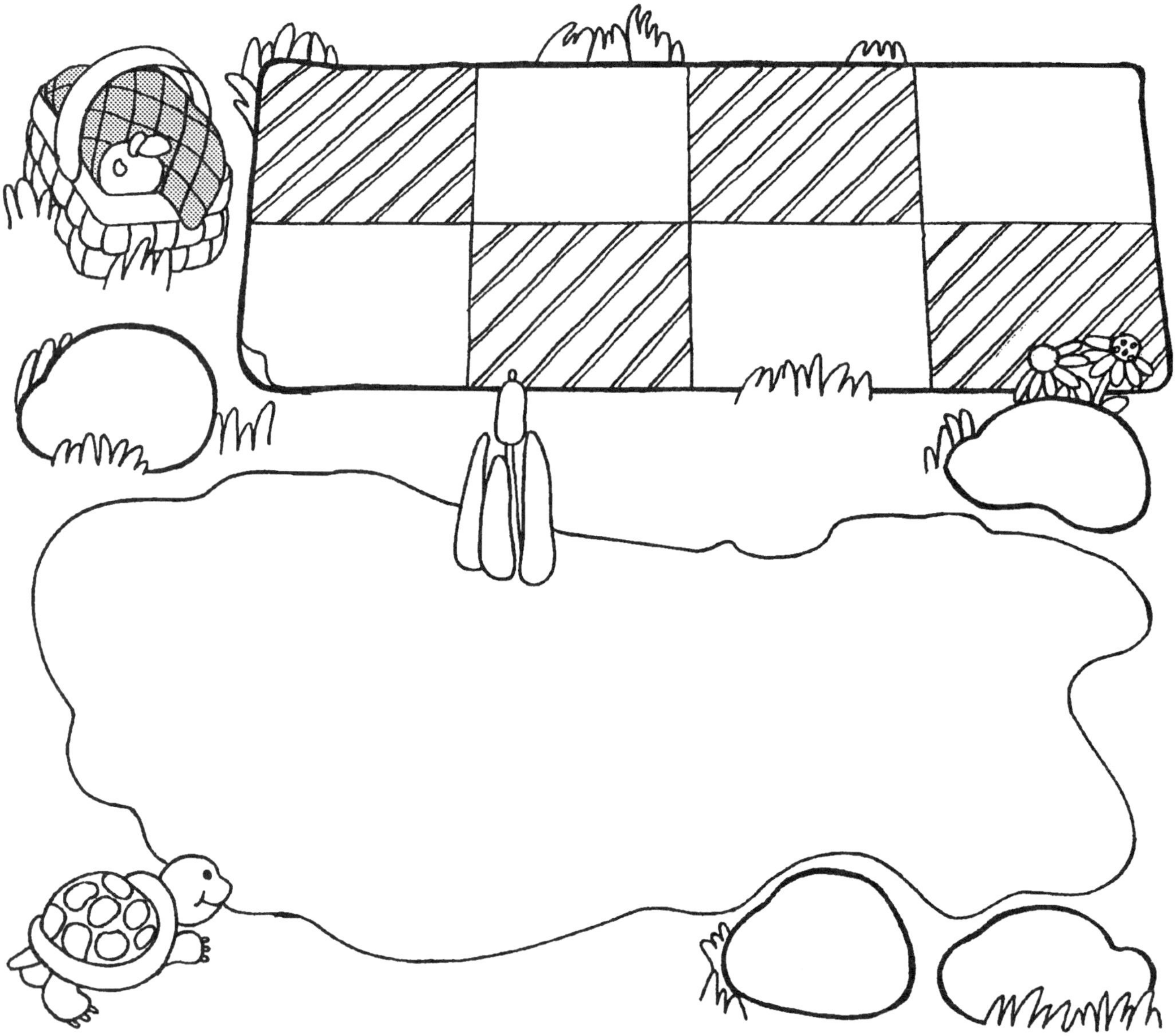

▶ **Next, play "Will it Fill?"**

Too Many, Not Enough, Just Right

NAME ______________________________

Pick some bears.
Count and write the number in the correct column.
Use the 5 or 10-frame to check.

Too Many	Not Enough	Just Right

▶ **Next, play "Bear Hunt."**

Compare and Race to the Bell

NAME ______________________________

Pick a card and write the number.
Compare and circle the one that is greater.
The player with the greater number moves one space.

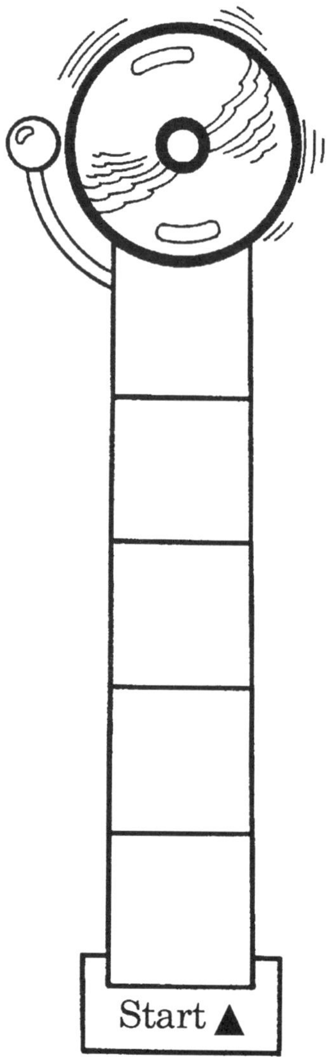

My Number	Your Number

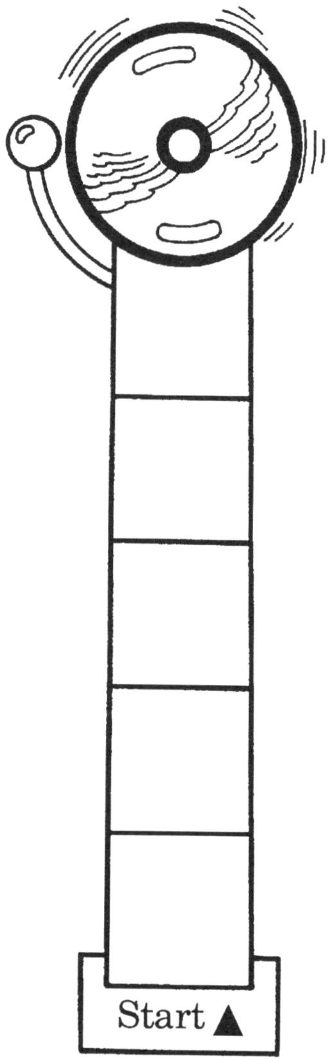

▶ **Next, play "Go for Less."**

Papa Says, Mama Says

NAME ______________________________

Listen to the bear story.
Act it out. Write the numbers.

Day 1	2 big handfuls	3 small handfuls
Day 2	1 big and 2 small handfuls	2 big and 1 small handful
Day 3	4 small handfuls	3 big handfuls

► **Next, do "More Berries."**

Deliver the Mail

NAME ____________________

Listen to the bear story.
Write numbers on the mailboxes.

		37
55		

	44	
	64	

77		
		99

16		
		38

		40

▶ **Next, play "Peep and Tell."**

Teaching Notes
Estimating and Measuring

Bear Totem Poles
(Non-Standard Units of Measure.)

Warm-Up: Distribute several bears to each pair of children. On the overhead or chalkboard place or tape several Baby Bears end to end to form a 6-bear totem pole. Have children copy and estimate how many Papa Bears are needed for a totem pole just as high, and use bears to check. Use different numbers of bears to compare the Baby totem poles with the Papa or Mama totem poles. (5 Papa Bears are about the same height as 6 babies; and 4 Mama Bears are about the same height as 5 babies.)

The Page: Provide Baby Bear counters or those from Teaching Aid 1 for each pair of children. Place 5 pieces of tape or identify 5 objects in the room to be measured. Label them A,B,C,D,E. Working in pairs, children first estimate and record how many bears long each is, then measure and record the actual number of bears.

More For You: Compare. Children repeat the activity using Papa Bears to measure. Encourage them to compare with the list on the page, then write or tell about their comparison.

Bear Jump

Warm-Up: Work with children to agree on a procedure for measuring the distance between two bears. Place two bears on the overhead projector (or tape them to the chalkboard). Place a finger by each bear and ask children to estimate how many Baby Bears could be placed between them. How many Mama Bears? How many Papa Bears?

The Page: Provide two bears to each pair of children. Taking turns with their partner, ask children to drop two Baby Bears on a table or desk. Have students first record their estimate of the distance between the bears, then measure the distance to check their estimate.

Suggest that children drop bears from the height of their partner's knee to complete the first table; from the height of their partner's shoulder for the second. Younger children can tell how many Baby Bears apart they landed. You may wish to have older children measure in centimeters or inches.

Summarize the activity by leading children to observe and discuss whether their estimates were better in the later trials; whether the distances from knee high were different than those from shoulder height, etc.

More For You: Write About It. Have children write about their observations from the activity. Ask volunteers to share and compare ideas.

Bear Totem Pole

NAME ______________________________

Write your estimate. Write your measure.

	My Estimate		My Measure
A.	____________		____________
B.	____________		____________
C.	____________		____________
D.	____________		____________
E.	____________		____________

▶ **Next, do "Compare."**

Bear Jump

NAME ____

> Drop two bears.
> Guess and write how far apart they will land.
> Measure and write.

How Far Apart?

Knee High

Drop	1	2	3	4	5	6
My Guess						
My Measure						

Shoulder High

Drop	1	2	3	4	5	6
My Guess						
My Measure						

▶ **Next, do "Write About It."**

Teaching Notes
Adding and Subtracting

Note to Teachers: This section contains 8 story mats for acting out many different kinds of addition and subtraction situations. Numbers suggested in the sample stories may be changed to fit student needs.

Assessment: For each situation, ask questions to determine what children understand.

- Do they recognize addition (subtraction) situations?
- Which techniques or thinking strategies do they use to solve problems?
- Do they use fingers or counters?
- Do they count on from the first addend? from the greater addend?
- Do they use known facts? What other solution strategies are used?
- Do they understand the connection between what is done and what is recorded?

Children should be encouraged to tell or write the matching number sentence in each situation.

Using the Problem Types: The problem types of this section are arranged in a progressive sequence based on what most children find "easy" vs. "hard." Because children grow in their solution approaches from using objects to more sophisticated thinking strategies, it is important to have a sense of how the children are thinking and to extend that thinking through cooperative interaction with partners and the larger group. Once children have shown they can handle a type of problem, it should be reused in story settings. Within a given session, several types of addition or subtraction situations can be presented — all based on children's experiences. Children should be encouraged to make up and tell stories of their own for others to solve.

Bear Tag
(Addition Stories with Action, Total Unknown)

Warm-Up: Involve children in acting out stories like the following:

"Today is the Bears' Sports Day. Let's go watch what they do. This part of the room will be our bus to take us on our trip. Anyone wearing red should come up first. Anyone wearing black come up next. How many are in our bus now?"

Unless addition is a very new experience, the number sentence that matches the story should be written by a child or the teacher on the chalkboard. Ask yourself the assessment questions (see Note to Teachers above), then continue with similar situations.

The Page: Children use bears (2 colors) to act out the story and write the number sentence to match.

"The first game the bears played at Sports Day was Bear Tag. To start, Papa Bear was 'It.' The big rock was 'Home.' He ran as fast as he could but 3 red bears beat him to the rock. He chased again, but 5 others got there before Papa. They sat down with the 3 red bears and just grinned at Papa. How many bears were at 'Home'?"

Children can think about their response and how it could be recorded, compare with their partner, and then share with the group. Have children record their answers in the space provided. Ask assessment questions. Tell other stories to continue the Sports Day theme, using numbers appropriate for the children's skills.

More For You: Tell a Story. Working with a partner, ask children to take turns creating a story for one of the number sentences on the page. The other child can identify the number sentence.

Bear Vote

(Different Ways to Split a Whole)

Warm-Up: Involve the children in acting out situations like the following:

"Think about it: if you had a choice between an apple or a banana (draw these on the chalkboard, 3 or more feet apart), **which would you choose?"**

Select 5 students to come up and line up near their choice. Write (or have a child write) the addition number sentence to match (e.g. 2 + 3 = 5). Select other groups of five and record the different ways that the students partitioned themselves. Ask if there are any other ways that this could have been done. Continue with similar situations.

The Page: Provide bear counters and tell stories like the following:

"The bears drew a picture of each flower they saw on their field trip. Which was their favorite: a tulip or a daisy? Teacher Bear asked the 8 bears to line up by their favorite flower. How could the bears have lined up?"

Have children tell the ways, then write them. Additional lines can be drawn as needed. Repeat for other numbers of bears.

More For You: All the Ways. Children select a number for a class of bears that is different from any number used on the page and try to find all the ways the bears could line up to vote. Before moving the bears, children might be challenged to predict how many ways there will be. Discuss patterns that the children observe.

Bear Porridge Hot

(Addition with No Action, Total Unknown)

Warm-Up: Give each pair of children a Papa Bear and some two-color counters. Have one child draw a loop for a bowl on plain paper and act out the story while the other writes to record what is done.

"'Bear porridge hot. Bear porridge cold. Bear porridge in the pot 9 days old.' One bright Sunday morning Papa Bear sang this rhyme as he fixed porridge for Mama. Papa Bear decided to put fruit on the porridge. When Mama sat down she saw 4 raspberries and 7 blueberries. How many berries did Mama Bear get?"

Continue with similar situations.

The Page: Provide two-color counters and tell stories which continue the theme above by relating what happened on Monday, on Tuesday, and so on. Children should act out and record the number sentence for each. Use number sentences that are appropriate to your student's skills.

More For You: Tell a Story. Students write numbers appropriate to their level (up to 9) on paper scraps and place under a book or in an envelope. Taking turns, one student draws 2 or 3 numbers and writes an addition sentence using them as addends. The other tells a story for the numbers. Together students decide on the total and use counters to check.

Bear Lake

(Take Away Subtraction)

Warm-Up: Involve children in acting out situations like the following: Start by selecting 7 children to come to the front of the room. Say **"This part of the room is the lake. Would some of you like to jump in for a swim?... How many did not jump in?"**

Each time, the number sentence that matches the story should be written by a child or the teacher on the chalkboard. Continue as time and interest permit.

The Page: Provide bear counters to pairs of children and tell stories like the following:

"On Sunday nine bears went to Bear Lake. Two bears decided to go in for a swim. How many did not go for a swim?"

Have children think about their response and how it could be recorded. Then have them compare with their partner and share with the group. Ask children to record their response in the space provided. Tell other stories, continuing the Bear Lake theme.

More For You: Tell a Story. Working with a partner, ask children to take turns creating a story for one of the number sentences on the page. Have the other child identify the number sentence.

Bear Picnic

(Comparison Subtraction: How Many More/Fewer Than?)

Warm-Up: Divide the class into two groups, and involve the children acting out situations like the following:

"It's clean-up time for this room. I've been thinking about all the things I'd really like to have done. I'd really like to have 2 people from Group 1 straighten the books on the bookcase and 5 people from Group 2 pick up paper scraps from the floor. How many more children from Group 2 helped me than Group 1?"

To help, children from the two groups can line up facing each other. Pay particular attention to solution strategies as you ask leading questions: **"How do you solve the problem? Do you count on, count back or take away?"** Continue the procedure using other volunteers.

The Page: Provide bear counters to pairs of children and tell stories like the following:

"The bears went on a picnic. They found a table just their size. 6 Baby Bears sat on one side of the table. 3 Papa Bears sat on the opposite side. How many more Baby Bears were there?"

Have children think about their response and how it could be recorded. Then have them compare with their partner and share with the group. When they have had some experience with "How many more?" situations, focus on the idea of "how many fewer."

More For You: Pick Some Bears. Have children work with a partner. Each takes some bears and lines them up. Ask them to take turns creating a story comparing the number of bears in each line.

Birthday Present for Bear

(Missing Addend)

Warm-Up: Start by calling 4 children to the front of the room. Involve them in acting out situations like the following: **"We have 4 children. We need 6 people for this game. How many more are needed?"** Repeat with several other examples.

The Page: Provide counters to pairs of children and tell stories like the following:

"The bears are getting ready for a birthday party for Baby Bear. Mama Bear is telling the other bears how to hang the sign. Baby Bear is playing with 5 toys. Papa Bear said, 'There are some toys in this present,too. When you open it, you will have 8 toys. How many toys do you think are in the box?'"

Have children think about their response and how it should be recorded. Ask them to compare with their partner and share with the group. Then tell other stories using the Bear Birthday Present theme. You may wish to vary the activity by presenting situations such as:

"Some bears came to the party early. Later, three more bears came. Then Mama Bear counted 9 bears at the party. How many bears were there early?"

More For You: Just Right, Not Enough or Too Much? Have students write numbers between 5 and 12 on paper scraps and turn them face down. One child draws a number while the other takes some bears. Taking turns, one child tells whether the number of bears drawn is just right, not enough or too much. A point is earned only when there are not enough and the player can tell how many more are needed.

Bear Hide-and-Seek

(Change Unknown)

Warm-Up: Start by calling 8 children to stand near you by the door and involve children in acting out situations like the following:

"Eight children are here. While the rest of you turn away and close your eyes, some of these children will go out in the hall (tap the shoulders of three children to leave). Look now. How many?"(5)

If appropriate, write the number sentence to match: 8 - ___ = 5. **"How many went out into the hall?"** (3) Repeat with different numbers of children.

The Page: Distribute some bears to each pair of children and tell stories like the following:

"The bears are playing a game of Hide-and-Seek at Baby Bear's birthday party. 10 bears want to play. Some bears ran quickly and hid in the cave. Four bears were still trying to decide where to hide. How many bears hid in the cave?"

Have students take turns, hiding some bears, and recording the actions. Together they decide on the number sentence to match, and tell the number of bears that are hidden.

More For You: How Many are Hidden? With a partner, children select and count some bears. While one looks away, the other child can hide some under a box or paper. The second child notes the number that can still be seen and earns one point for telling the correct number of hidden bears. The winner is the child with the most points when time is up. (Note: Some children may need to use bears to solve the problem.)

Bear Tag

NAME ________________________________

Listen to the story. Use bears to act it out. Write the number sentence.

+ ____________________ ____________________

____________________ ____________________

▶ **Next, do "Tell a Story."**

Bear Vote

NAME ____

Listen to the story.
Use bears to act it out.
Write the number sentence.

▶ **Next, do "All the Ways."**

Bear Porridge Hot

NAME ______________________________

Listen to the story.
Use counters to act it out.
Write the number sentence.

______________________ ______________________

______________________ ______________________

▶ **Next, do "Tell a Story."**

Bear Lake

NAME ____

Listen to the story.
Use bears to act it out.
Write the number sentence.

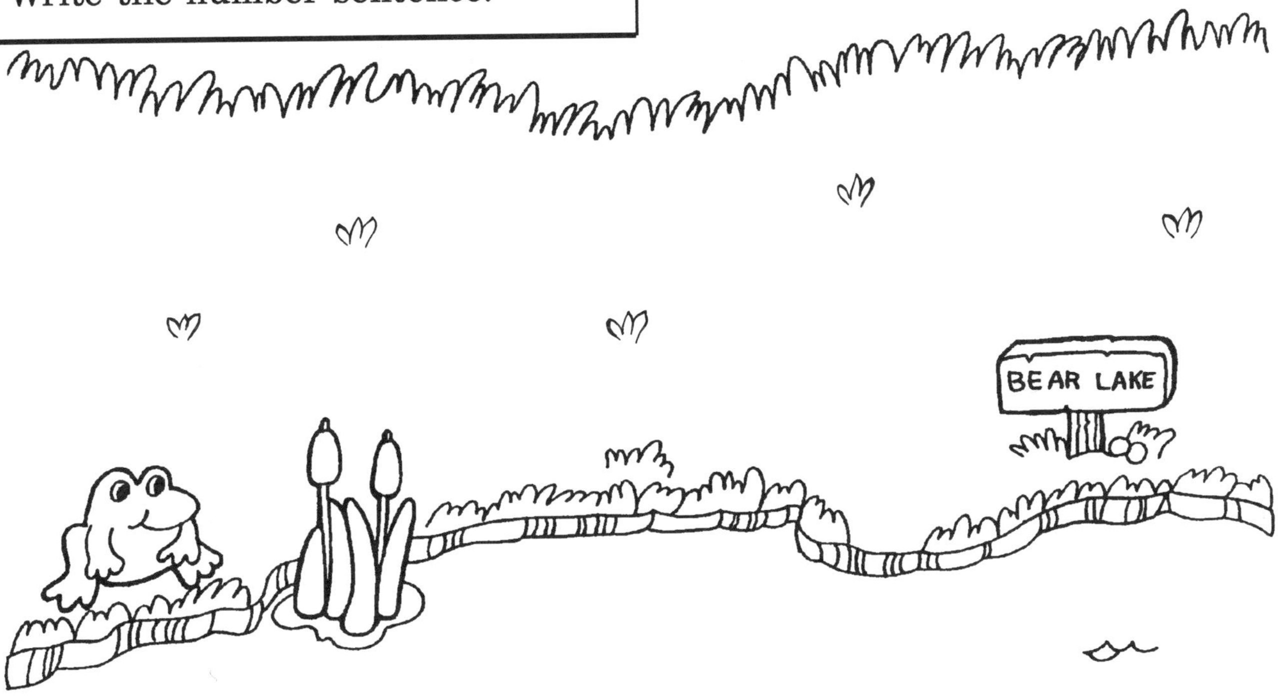

______________________ ______________________

______________________ ______________________

▶ **Next, do "Tell a Story."**

Bear Picnic

NAME ______________________________

Listen to the story.
Use bears to act it out.

► **Next, do "Pick Some Bears."**

Birthday Present for Bear

NAME ____

Listen to the story.
Use counters to act it out.
Write the number sentence.

______________________ ______________________

______________________ ______________________

▶ **Next, play "Just Right, Not Enough, or Too Much?"**

Bear Hide-and-Seek

NAME ______________________________

Listen to the story.
Use the bears to act it out.
Write the number sentence to match.

► Next, play "How Many are Hidden?"

Teaching Notes
Multiplying and Dividing

Note to Teachers: This section contains 8 story mats for acting out many different kinds of multiplication and division situations. Numbers suggested in the sample stories should be altered to fit student needs. Younger children can profit from acting out multiplication and division stories even though they may not record the number sentence to match. In addition to the more standard types of multiplication and division situations, comparison multiplication and division are also presented.

Assessment: For each situation, ask questions to determine what children understand.

- Do they recognize multiplication (division) situations?
- Which techniques or thinking strategies are used to solve the problem?
- Do they use fingers or counters?
- Do they count or use a counting sequence based on multiples?
- Do they add or subtract from a known product?
- Do they use known facts in other ways?
- What other solution strategies are used?
- Do they understand the connection between what is done and what is recorded?

Children should be encouraged to tell or write the matching number sentence.

Using the Problem Types: This section contains several types of multiplication and division problems. Because children grow in their solution approaches, it is important to have a sense of how children are thinking and to extend that thinking through cooperative interaction with partners and the larger group. Once children have shown they can handle a type of problem, it should be reused in story settings. Within a given session, several types of multiplication and division situations can be presented — all based on children's experiences. Students should be encouraged to make up stories of their own for others to solve.

Bears in the Band
(Arrays — Same Number in Each Row)

Warm-Up: Distribute 12 bears to pairs of students. Ask, **"How could you line up these 12 bears in rows so there are the same number in each row?"** Children work with their partner and then describe their way to others in the class. Each time, have children show their way on the overhead. Keep a chalkboard list of all the different ways: "2 rows of 6, 2 x 6 = 12."

The Page: Have children use bears to act out this story and write the number sentence to match: **"The Bears' Marching Band is about to go on the field. The flag bears are ready: 2 rows of four bears. How many bears are in the flag corps?"**

Continue the story for other sections of the band. Use other numbers of rows and bears in each row. Discuss how children determined the total number of bears. Encourage divergent thinking by asking, **"Did anyone think of a different way?"**

More For You: Some Do, Some Don't. Taking turns with a partner, one child takes some bears. Together they try to arrange them in arrays. Help students record arrays by writing a matching multiplication sentence. If children can, have them try to make a different array with the same number of bears. Encourage students to use different numbers of bears.

Honey Pots
(Groups of the Same Size)

Warm-Up: Involve the children in acting out stories like the following:

"Today we will play 'Go to the Corner'. At my signal, each of you should pick three other children and go to a corner of the room before I say, 'One corner, two corners, three corners, four corners.' Only one group is allowed at each corner. Ready? Go! ... Look and tell me what you see." (2 groups of three: 6 children. 2 times a group of 3 — that's 6 in all; 2 x 3 = 6.)

Continue the game with other numbers of corners and children at each corner.

The Page: Distribute some bears and scissors to each pair of children and have them cut out honey pots on the page. Children use the bears and the honey pots to act out each story and write the number sentence to match.

"All the bears in the Honey Pot Club loved honey. When they had their meetings, the same number of bears gathered around each pot. On Monday there were 4 pots of honey. Two bears sat beside each pot. How many bears attended the Monday meeting?"

Continue telling similar stories for other days.

More For You: Show the Story. Working with a partner, children make up a Honey Pot multiplication story. Then they can paste bears (Teaching Aid 1) and honey pots on a large piece of paper to illustrate it. Under the paste-up, children should write the story and the number sentence.

Bear Sing Along
(Multiplication Turnarounds)

Warm-Up: Place 2 rows of 3 Baby Bears on the overhead projector. Place Papa Bear facing the Baby Bears and ask children to describe what Papa Bear sees. (2 rows of three, 6 bears). Place Mama Bear to the side and ask children to describe what Mama Bear sees. (3 rows of two, 6 bears). Ask, **"What number sentences describe what the bears see? How are the number sentences alike? (Same numbers) How are they different?"** (Order is different) Repeat with different arrangements of bears.

The Page: Distribute some bears to each pair of children. Working together, they use the bears to act out the following story and write the two number sentences.

"The bears are almost ready for the Sing Along. The music teacher led the first group of bears to the risers. Mama Bear was sitting in front and she saw 2 rows of 5. Papa Bear was sitting on the side and he saw 5 rows of 2. Who saw more bears?" (Mama and Papa Bear saw the same number of bears — just turned differently. 2 fives is as many as 5 twos. 2 x 5 = 10 and 5 x 2 = 10.)

Tell children that turnarounds in multiplication have the same answer. Continue the Sing Along theme for other groups.

More For You: Show a Story. Working with a partner, have children make up a Sing Along multiplication story and paste bears (Teaching Aid 1) on the page to illustrate it. Under the paste-up, children should write the number sentences for the story.

Fishing at the Lake

(Comparison Multiplication)

Warm-Up: Distribute some bears and have children act out comparison multiplication situations. Display 3 bears on the overhead and say, **"Three bears are here. I want another group with twice as many bears."** Have children use bears to decide how many are needed, then tell a partner and share with the larger group.

When checking, call children's attention to the first group of bears on the overhead by saying, **"I have a group of 3 bears one time here."** Then refer to the second group of bears. **"Here I have a group of 3 bears one time, two times — so I have twice as many bears or 6 bears in this second group as in the first: 2 x 3 = 6."** Repeat with different numbers.

The Page: Distribute some bears and counters for fish to each pair of children. Working together, have them listen to the story and use bears and counters to act it out.

"Two Papa Bears had fished most of the afternoon. One Papa Bear said to the other, 'I have 2 fish. You have 3 times as many as I have.' How many fish did the other Papa Bear catch?"

Observe as children place 2 counters for the first bear's fish, then repeat that placement 3 times on the other side of the path for the second bear's fish. After children write the number sentence and complete it, they can refer back to it as they verbally describe the multiplication situation. Repeat for other numbers of fish.

More For You: Tell a Story. Working with a partner, have students take turns creating a comparison story for one of the multiplication sentences on the page. The other child identifies the number sentence, using bears and "fish" counters to help or to check thinking. Provide large paper so children can paste bears (Teaching Aid 1) and draw fish to illustrate one of the stories. Under the paste-up, you may wish to have children write their story and the matching multiplication sentence.

Bear Photo Day

(Missing Factor)

Warm-Up: Provide bears so children can act out stories like the following:

"All the bears brought their cameras to Family Reunion Day. Papa Green Bear gathered the first family to take their photo. He seated them in 3 rows, 3 bears in each row. How many bears were in the first family?"

Repeat with different numbers of bears.

The Page: Distribute some bears to each pair of children. To start, write this number sentence on the chalkboard: 3 x ☐ = 9. Working together, have students listen to the story and use the bears to help find the missing numbers.

"The day after the Family Reunion Day, the bears sent their film to be developed. There were 9 pictures in all, and 3 bears shared them equally. How many pictures did each bear get?"

Use the same scenario to create other number sentences.

More For You: Under My Thumb. Working together, have groups of children make arrays of bears and write the multiplication number sentence for each on a separate strip of paper. After 6 sentences are written, children turn them face down. Taking turns, one child draws a number sentence and covers one factor with a thumb. The other tells the number that is missing, using bears either to determine the number or to check that it is correct.

Water Play

(Division: Total and Number of Groups Known — Sharing)

Warm-Up: Use bears on the overhead projector and involve children in acting out "sharing" situations. Start by placing a handful of bears on the overhead for all to see. Ask two volunteers to come to the overhead and say, **"These two children will share the bears. Each will get the same number. Think and tell your partner how many bears each will get."** Ask the children at the overhead to decide between themselves and take their bears. Discuss alternatives on what to do with any extras. Repeat with different numbers of bears, recording each expression on the chalkboard.

The Page: Have children cut apart the water rafts of Teaching Aid 9 and distribute some bears to each pair of children. Working together, they listen to the story and put bears on the rafts to act it out.

"It was a great day and the bear families were glad to be at the lake. Mama Bear looked at the 12 bears ready to go in and threw 4 rafts into the water. There are enough rafts for everyone. Help the bears share. Put the same number of bears on each raft."

Have the children place bears on the rafts "in" the water to help or to check thinking.

More for You: Tell A Story. Provide large paper so children can paste bears (Teaching Aid 1) onto rafts to model one of the stories. Under the paste-up, have children write their story and the matching division sentence.

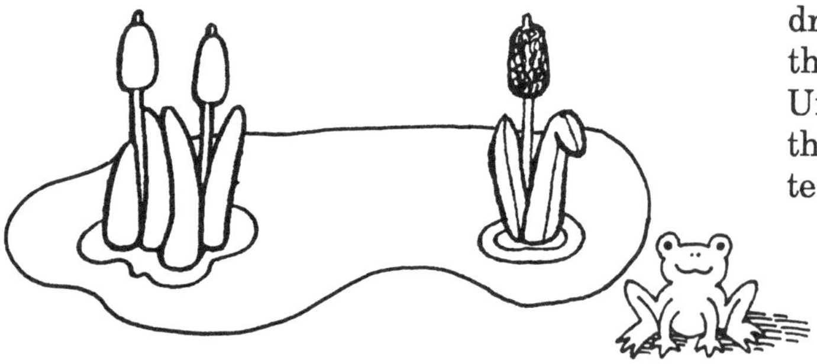

Bear Camp Out

(Division: Total and Number in Each Group Known)

Warm-Up: Use bears on the overhead projector and involve children in acting out division situations in which the number of groups is unknown. Say: **"I want to give two bears to every person I call up here. I want to give out all the bears. Think and tell your partner: how many children do you think I can call on?"** Call on children to share their thinking, then have children come up to "act out" what has been decided. Repeat with different numbers.

The Page: Have children cut apart the pup tents on Teaching Aid 9 and distribute some bears to each pair of children. Working together, they listen to the story and put bears "in" the tents to act it out. Then they write the matching division sentence.

"The bears hiked to the campground. Papa Bear was to give out the tents. He looked at the 10 bears and decided that 2 to a tent was enough. How many tents were needed for the bears?"

Encourage children to place bears "in" the tents around the campground to help or to check thinking. Children may also write a number sentence to verbally describe the division situation. Repeat for other numbers of bears and tents.

More For You: Tell a Story. Working with a partner, children can take turns creating a story for one of the division sentences on the page. The other child identifies the number sentence, using bears and tents to help or check thinking. Provide large paper so children can paste bears (Teaching Aid 1) "in" the tents to illustrate one of the stories. Under the paste-up, children should write their story and the matching division sentence.

Bears in the Band

NAME ______________________________

Listen to the story.
Use bears to act it out.
Write about it.

______________________________ ______________________________

______________________________ ______________________________

► **Next, do "Some Do, Some Don't."**

Honey Pots

NAME ______________________________

Listen to the story.
Use bears to act it out.
Write about it.

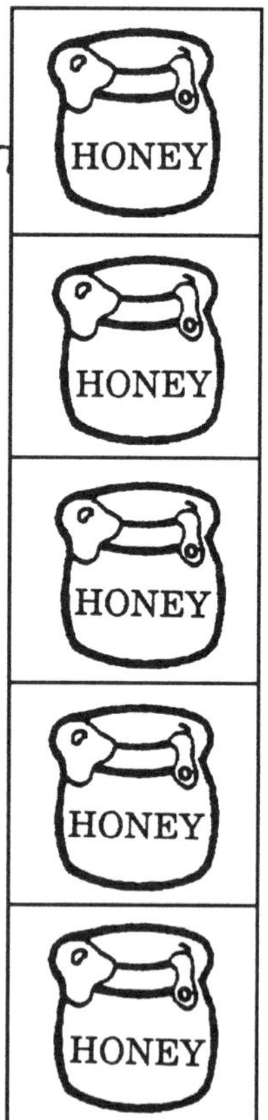

X ____________________ ____________________

____________________ ____________________

► **Next, do "Show the Story."**

Bear Sing Along

NAME ______________________________

Listen to the story.
Use bears to act it out.

X ______________________ X ______________________

▶ **Next, do "Show a Story."**

Fishing at the Lake

NAME ______________________________

> Listen to the story.
> Use counters to act it out.
> Write the number sentence.

______________________ ______________________

______________________ ______________________

▶ **Next, do "Tell a Story."**

Bear Photo Day

NAME ____________________

Listen to the story.
Use bears to help find the missing number.

x ☐ = ____________________ ☐ x ____________________

☐ x ____________________ x ☐ = ____________________

▶ **Next, do "Under My Thumb."**

Water Play

NAME ______________________________

Listen to the story.
Use bears and rafts to act it out.
Write about it.

______________________ ______________________

______________________ ______________________

▶ Next, do "Tell a Story."

Bear Camp Out

Warm-Up

NAME ______________________________

Listen to the story.
Use bears and pup tents to act it out.
Write about it.

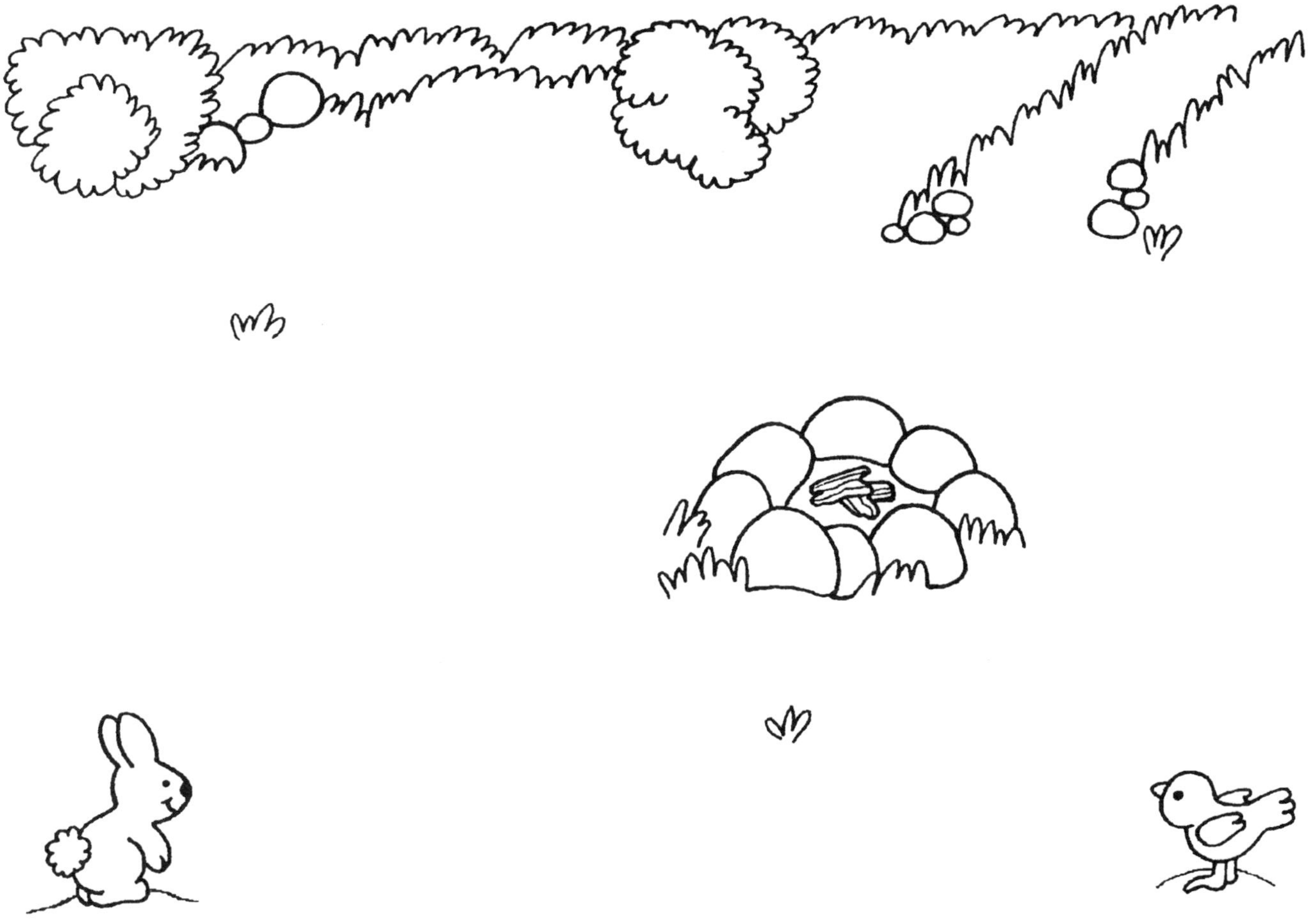

______________________ ______________________

______________________ ______________________

▶ **Next, do "Tell a Story."**

More for You

Teaching Notes
Fractions

Parts Are Small

Warm-Up: Place 1 Papa Bear and 2 Baby Bears on the overhead. Ask children what fractions tell about the bears. When sharing with the larger group, be sure all ways are described and written. (e.g., 1/3 are large bears, 2/3 are small bears). Emphasize language: **"1 out of 3 bears is large, so 1/3 of the bears is large."** If appropriate, call on a child to write the fraction to match each description.

Label each fraction: $\frac{1}{3}$ of the bears is large.

Repeat the activity for groups of 2, 4, 6, and 8 bears.

The Page: Distribute some Papa Bears and Baby Bears in a sack or box to each pair of children, as well as bears from Teaching Aid 1. To start, have one child pick 3 bears. Working together, the children should agree on the fraction that describes the part that is small. Have them write and paste the fraction using bears from Teaching Aid 1.

More For You: Many Ways. Independently, have children color each group of bears (two colors). Taking turns with a partner, a point is earned for every way that a group of bears is described using a fraction (e.g., 1/3 is red; 2/3 are yellow; 2/3 are large; 1/3 is small).

Half — More or Less

Warm-Up: Distribute bears of 2 colors (red, blue) — one bear to each child. Select four volunteers to come, with their bears, to the front of the room. Elicit the number of red bears and whether that number is half of the total, more than half or less than half. Call on a child to write the fraction that is represented. Repeat the activity with the other children.

The Page: Distribute some bears of 2 colors in a sack or box to each pair of children. To start, one child picks a bear; the other writes in the color (e.g., red). Then, taking turns, children take 4 bears; tell the fraction that is red and whether that fraction is half, more than half or less than half of the total number of bears taken. Partner records by writing the fraction in the appropriate section. Repeat for other fractional parts.

More For You: Pick and Show. Taking turns, one child points to a fraction on the page. The other uses bears to illustrate it. Children should verbally describe what they have shown.

Half-Ways

Warm-Up: Call four children to the front of the room and have two of them sit down. Have children describe to a partner what was done. When sharing with the larger group, have children focus on different ways to verbally express the same situation: "2 out of the 4 sat." "Half sat." If appropriate, call on a child to write the fraction to match the descriptions.

Label each fraction:

$\frac{2}{4}$ of the children sat.

Also note that half remained standing. Repeat the activity for groups of 6 and 8 children. If appropriate, also pose the following problem: **"Suppose I had a group of 5 students. Could half sit down?"**

The Page: Distribute some bears of 2 sizes and bearprints from Teaching Aid 1 to each pair of children. Working together, ask children to line up the number of bears specified, making sure each time that half are larger. Help students agree on the fraction that describes the part that is larger (e.g., 2/4, 3/6, 4/8); write it and paste in the bear prints to match (Teaching Aid 1). **Note:** When wrapping up the activity, emphasize the fact that there are many ways to show half.

More For You: Many Ways. Distribute bearprints from Teaching Aid 2. Independently children color each group (Papa, Mama and Baby sizes) of bearprints two colors. Taking turns with a partner, a point is earned for every way that a group of bears is described using a fraction (e.g., 1/4 is red; 3/4 are yellow; 1/2 is large, so 1/2 is small).

Parts are Small

NAME ______________________________

> Pick 3 bears.
> Tell the fraction for the part that is small.
> Write the fraction and paste the bears.

☐ of the
———
☐ bears are small

☐ of the
———
☐ bears are small

☐ of the
———
☐ bears are small

▶ **Next, do "Many Ways."**

Half — More or Less

NAME ______________________________

Use 2 colors of bears.
Pick one bear and write the color. Take turns.
Tell your partner the fraction and where to write it.

The color is ____________________ .

More than half	Half	Less than half

▶ **Next, do "Pick and Show."**

Half–Ways

NAME ______________________________

> Line up bears so half are larger.
> Write the fraction.
> Paste bearprints to match.

1. Use 4 bears.

$\frac{\square}{\square}$ of the bears are large

---

2. Use 6 bears.

$\frac{\square}{\square}$ of the bears are large

---

3. Use 8 bears.

$\frac{\square}{\square}$ of the bears are large

▶ **Next, do "Many Ways."**

Teaching Notes
Problem-Solving

Bear Tug
(Act It Out; Use Logical Reasoning)

Warm-Up: Call four volunteers to the front of the room. Give each child a blue (B), yellow (Y), red (R) or green (G) bear. Then give the following directions and have the children position themselves accordingly: **"Red is first. Blue is last. Green is in front of yellow."** (RGYB) Repeat with other volunteers using different directions.

The Page: Children use bears for the story as it is read:

"The Bear Tug was ready to start. Team 1 lined up this way: G was in front of R. Y was between R and B. (GRYB) For Team 2, Y was last. R was between Y and G. G was behind B." (BGRY)

Children color to record.

More For You: Match My Bears. With a visual barrier between two children, one child arranges bears and gives clues. The other arranges bears to match.

Bear Family Get Together
(Act It Out; Using Logical Reasoning)

Warm-Up: Children get 3 colors of bears, one of each size for a total of 9 bears. Give directions for placing bears in rows and columns on the page: **"Arrange the bears so each row has the same color; so each column has the same size bear.** OR **Put a Mama Bear in the 2nd row, 2nd column. Put a Papa Bear in the 3rd row, 1st column..."** Repeat the activity if needed.

The Page: Children use the bears for the story as it is read:

"The 3 bear families watched a video. When they looked around, they noticed something special: each row of bears was the same color; each column of bears had a bear of each size. How were the bears arranged?"

Have students cut and paste bears from Teaching Aid 1 to record.

More For You: Bear Mix-Up. Ask children to do the following activities: (1) arrange 3 colors of bears so each row and each column have a different color. (2) arrange 4 colors of bears so each row and column have a different color. (3) arrange 4 colors of bears so each row, column, and diagonal have a different color.

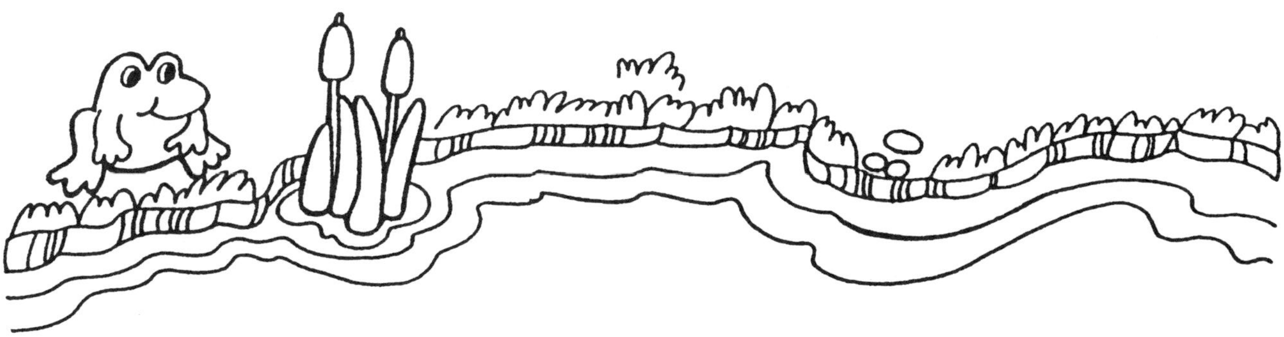

Bears for Sale

(Making an Organized List)

Warm-Up: Write bear prices on the board: Papa - 6¢; Mama - 4¢; Baby -2¢. Have children work with their partner to find out which costs more; how much 2 Baby Bears would cost; how much 1 Mama and 1 Papa would cost, and so on. Each time, call on children to share their responses with the larger group.

The Page: Encourage cooperative problem-solving. Provide pennies and hints as needed, to guide children in making an organized list. What would 3 Baby Bears cost? 2 Baby Bears and 1 Mama? 1 Baby Bear and 2 Mamas? Encourage children to write about the least or most cost, and discover that the cost could only be an even number.

More For You: What Can You Buy? Provide cent cards 2¢, 4¢, 6¢,...20¢. Children draw a card and earn 1 point for each way they can find to buy bears (any number) with the value drawn.

Spin and Buy

(Add; Use Logical Reasoning)

Warm-Up: Write bear prices on the board (or use overhead bears with price tags): yellow - 6¢; green - 3¢; blue - 2¢; red - 1¢. Have children tell their partner how much 2 blue bears would cost; how much 1 red bear and 1 green bear would cost, and so on. Each time, call on children to share their answers with the larger group.

The Page: Show how to place the pencil point on the spinner dot and to spin the paper clip. Ask children to color the bears and then follow directions to solve the problem.

More For You: Which Costs More? Children each pick up a handful of bears, estimate which group costs more and calculate (mentally, with calculator or paper-pencil) the total cost of each group.

Bear Balance

(Act It Out; Complete a Table; Look for a Pattern)

Warm-Up: Using a balance scale,* place 3 Baby Bears on one end, and ask students how many Baby Bears it will take to balance the scale. Repeat for other numbers of like bears. If children have no difficulty balancing Baby Bears against Baby Bears, then balance Baby Bears against Mama or Papa Bears; or Papa against Mama Bears. (Teacher information: Papa Bear = 12g; Mama = 8g; Baby =4g.

*Use Learning Resources Beginner's Balance (LER 740) or make one by hanging containers from a coat hanger.

The Page: Children listen to the story below, then finish the table to solve the problem. Children then write up their solution. If children have any difficulty completing the table, provide access to a balance scales.

"One sunny day, Green Mama Bear took 2 Baby Bears to the park to play on the teeter totter. Mama Bear sat on one end of the teeter totter and Green Baby Bear sat on the other. Mama Bear was heavier, so her side of the teeter totter was lower. Red Baby Bear decided to join in the fun. When he sat down with Green Baby Bear both sides of the teeter totter were even. The 2 Baby Bears weighed the same as 1 Mama Bear! How many Baby Bears would it take to balance 2 Mama Bears? 3 Mama Bears? .... 6 Mama Bears?"

More For You: Balance Papa and Baby Bears. Children complete the table to tell the number of Baby Bears necessary to balance different numbers of Papa Bears. If appropriate, provide paper so children can write about what they find.

Bear Tug

NAME ____

Listen to the story.
Place the bears on the page for the game.
Color to record.

▶ **Next, play "Match My Bears."**

Bear Family Get-Together

NAME ____

Listen to the story. Act it out with bears. Color to record.

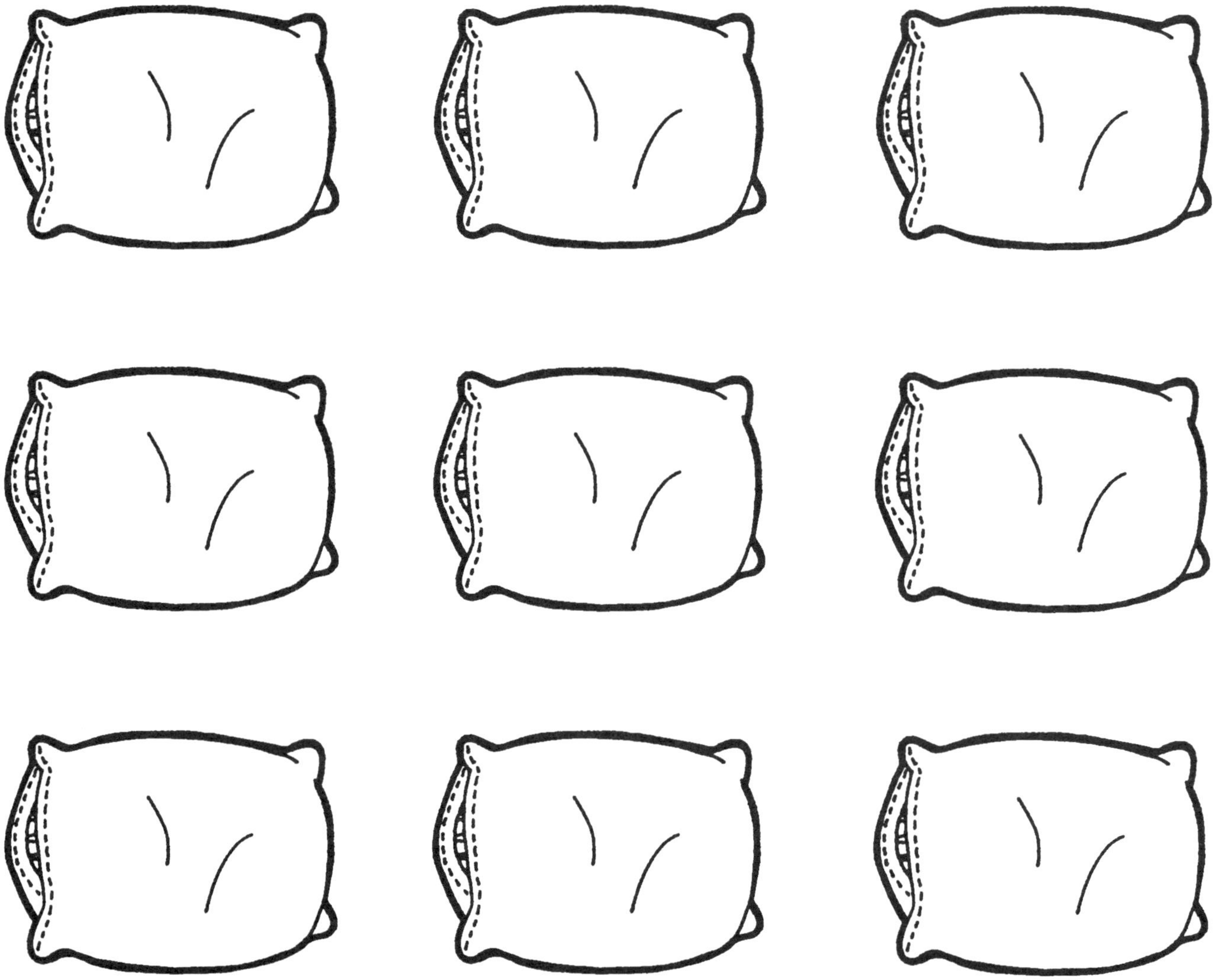

▶ **Next, do "Bear Mix-Up."**

Bears for Sale

NAME ______________________________

Jo wants 3 bears. Circle what they could cost.
Write what you learned.

1¢ 2¢ 3¢ 4¢ 5¢ 6¢ 7¢ 8¢ 9¢ 10¢ 11¢ 12¢ 13¢ 14¢ 15¢ 16¢ 17¢ 18¢ 19¢ 20¢

▶ **Next, play "What Can You Buy?"**

Spin and Buy

NAME ______________________________

Color:

Spin. Take that many bears.
Mark the total cost.
What numbers cannot be marked?

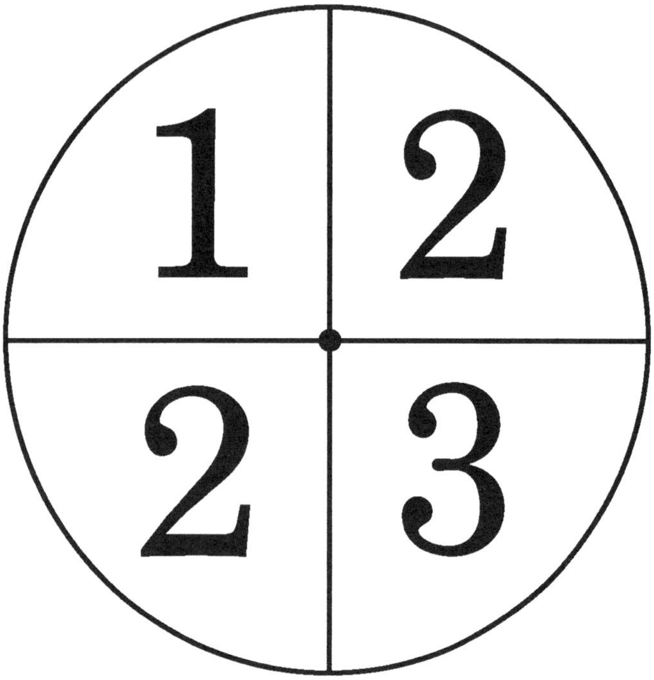

1¢ 2¢ 3¢

4¢ 5¢ 6¢

7¢ 8¢ 9¢

10¢ 11¢ 12¢

13¢ 14¢ 15¢ 16¢ 17¢ 18¢

▶ Next, play "Which Costs More?"

Bears Balance

NAME ______________________________

Listen to the story.
Finish the table to solve the problem.
Write about it.

Number of Bears to Balance

Mama	1	2				
Baby	2	4				

____

____

____

▶ **Next, do "Balance Papa and Baby Bears."**

Papa	1					
Baby	3					

Teaching Notes
Graphing and Probability

The Empty Honey Pot

Warm-Up: Tape a Papa, a Mama and a Baby Bear to the chalkboard. Ask three volunteers to draw one line for each bear as the following story is read:

"Mama Bear wondered why the honey pot was empty so soon. She gathered the family together to ask about it. Everyone told what they had eaten. Papa Bear began by saying, 'I remember seeing the pot full of honey when I got up this morning. I put 3 scoops on my porridge. And after breakfast I took just two more.' Mama Bear recalled, 'I took 2 scoops while I was waiting for the porridge to cook, and had another for a morning snack.' Baby Bear said, 'After my nap I had a one-scoop honey snack at 3, 4, and 5 o'clock. After dinner it was bedtime, so I didn't eat any more.'"

Get students to summarize the data by telling the total number of scoops for each bear. Ask, **"Why do you think the pot was empty?"**

The Page: Reread the story from the Warm-Up and help children finish the bar graph. Have children write in the missing numbers and suggest a title for the graph. Encourage children to think about the graph, to talk it over with a partner, and to write about it on the lines provided. Provide time for group sharing. Ask questions such as: **"Did the graph come out the way you thought it would? Who had the most scoops? the least? Do you think the results would be the same if the bears had this talk on a different day? Why do you think so?"**

More For You: Does Not Tell. Have students work with a partner and list one or more things the graph does not tell about the honey pot (its size; the size of the scoops that were taken...).

Two-Hand Scoop

Warm-Up: One at a time, have four students take a two-hand scoop of bears from a box; estimate and then count the number. Keep a list of results on the board.

The Page: Provide copies of Teaching Aid 8. Students use data from the Warm-Up to finish and analyze a double bar graph, comparing the estimate to the actual count for the four children. Ask children to write in the missing numbers, label the graph (write in the four children's initials) and give the graph a title. Have children color code the key (e.g., guess = yellow; count = green). Finally, have students color, cut out and paste strips to match the data.

Encourage them to think about the graph, talk it over with a partner, and then to share their ideas with the class. Provide time for group sharing. Discuss questions such as: **"Who came close to their guess? Do you think the results would be same if we did it again? Why do you think so? Did the last student guess "better" than the others? Why do you think so? (better basis for comparison.) Who had the greatest count? Is there a reason for this?"**

More For You: Does Not Tell. Students work with a partner and decide on one or more things the graph does not tell about the scoops that were taken. (the color or size of the bear that was drawn; how many of each kind, etc.)

Which Way, Bear?

Warm-Up: Provide some bears and a paper clip for each pair of children. Ask them to color the spinner on the page using two colors. If necessary, remind children how to place a pencil point on the dot and spin a paper clip for the spinner.

Taking several turns with a partner, have children spin and take a bear to match the color of the spin. After several spins, encourage children to share results (the color with the most space on the spinner came up more often).

The Page: Have children color the stars under the honey pot/bear's den to match the colors on the spinner. To start, children place a Baby Bear on the footprint and predict where Baby Bear will end up: eating honey or cleaning the den. Then taking turns, children spin and move one space toward the color of the spin. Continue until the honey pot or the bear den is reached. If time allows, encourage children to play again to see if the result is the same.

Compare results and ask questions such as: **"Where did your bear end up? What color did you color the big part of the spinner? What color did you color the smaller part? Why do you think Baby Bear got to the _____ first? If you played the game again, do you think the result would be the same? If you want Baby Bear to have a good chance of getting to the honey, how should the star by the honey and the spinner parts be colored?"**

More For You: Fair, Not Fair. Together discuss whether the game was fair or not fair. (The spinner is biased.) Have children suggest ways the game could be made "fair."

The Empty Honey Pot

NAME ______________________________

Use information from the story to finish the bar graph.
Tell your partner about the graph. Write about it.

Papa													
Mama													
Baby													
	1	2	3	4									

▶ Next, do "Does Not Tell."

Two-Hand Scoops

NAME ____

Use the information to finish the graph.

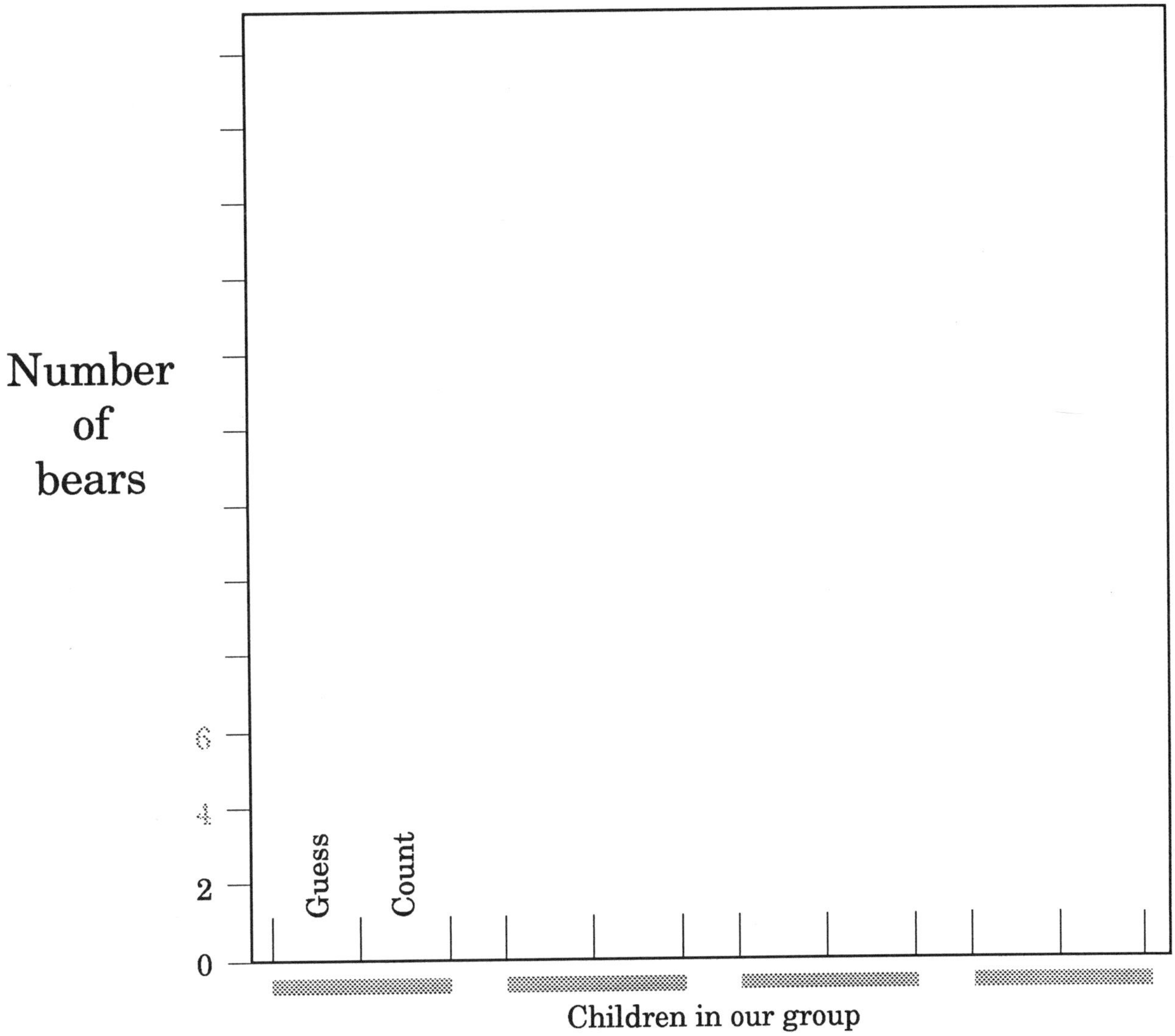

▶ **Next, play "Does Not Tell."**

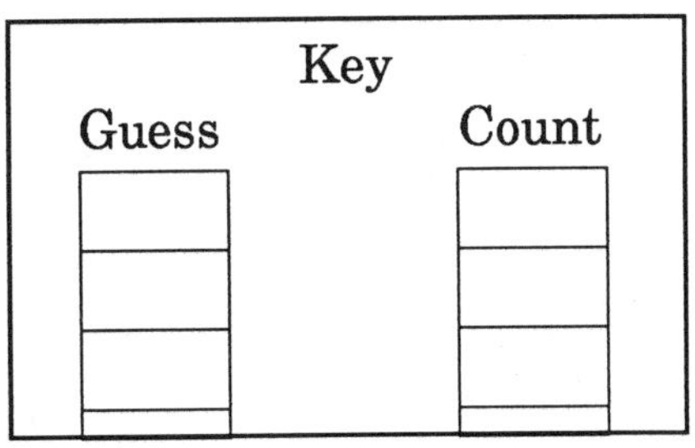

Which Way, Bear?

NAME ______________________________

Color the stars to match the spinner.
Put a Baby Bear on the .
Spin and move the Baby Bear.

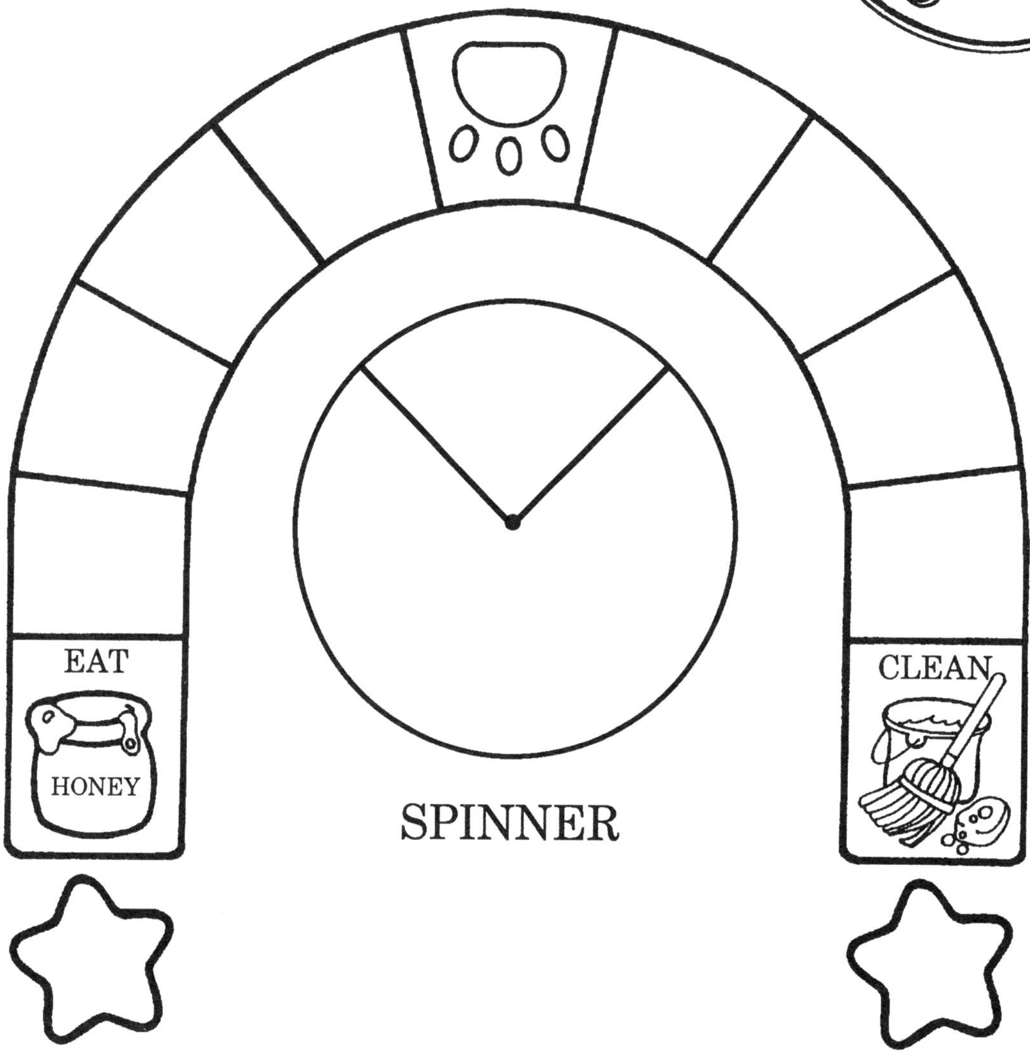

▶ Next, do "Fair, Not Fair."

Teaching Aid 1: Cut-Out Bears

Teaching Aid 2: Bearprints

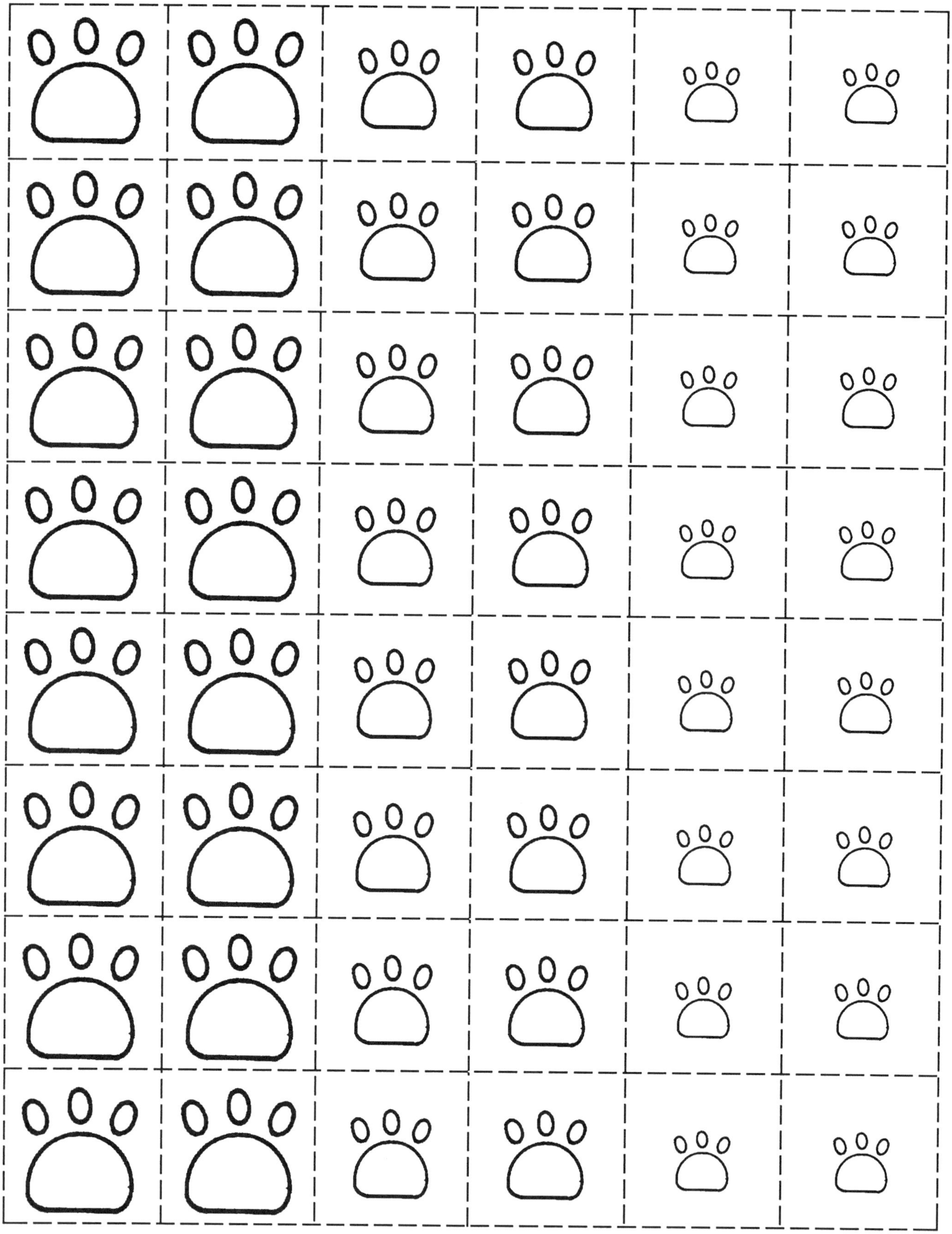

Teaching Aid 3: Counting Frames & Teddy Gram

5-Frame

10-Frame

Teddy-Gram

Dear Family,

Today is a BEARY special day because

____

____

____

____

____

Date ________________ Teacher ________________

Teaching Aid 4: Bear Hunt Gameboard

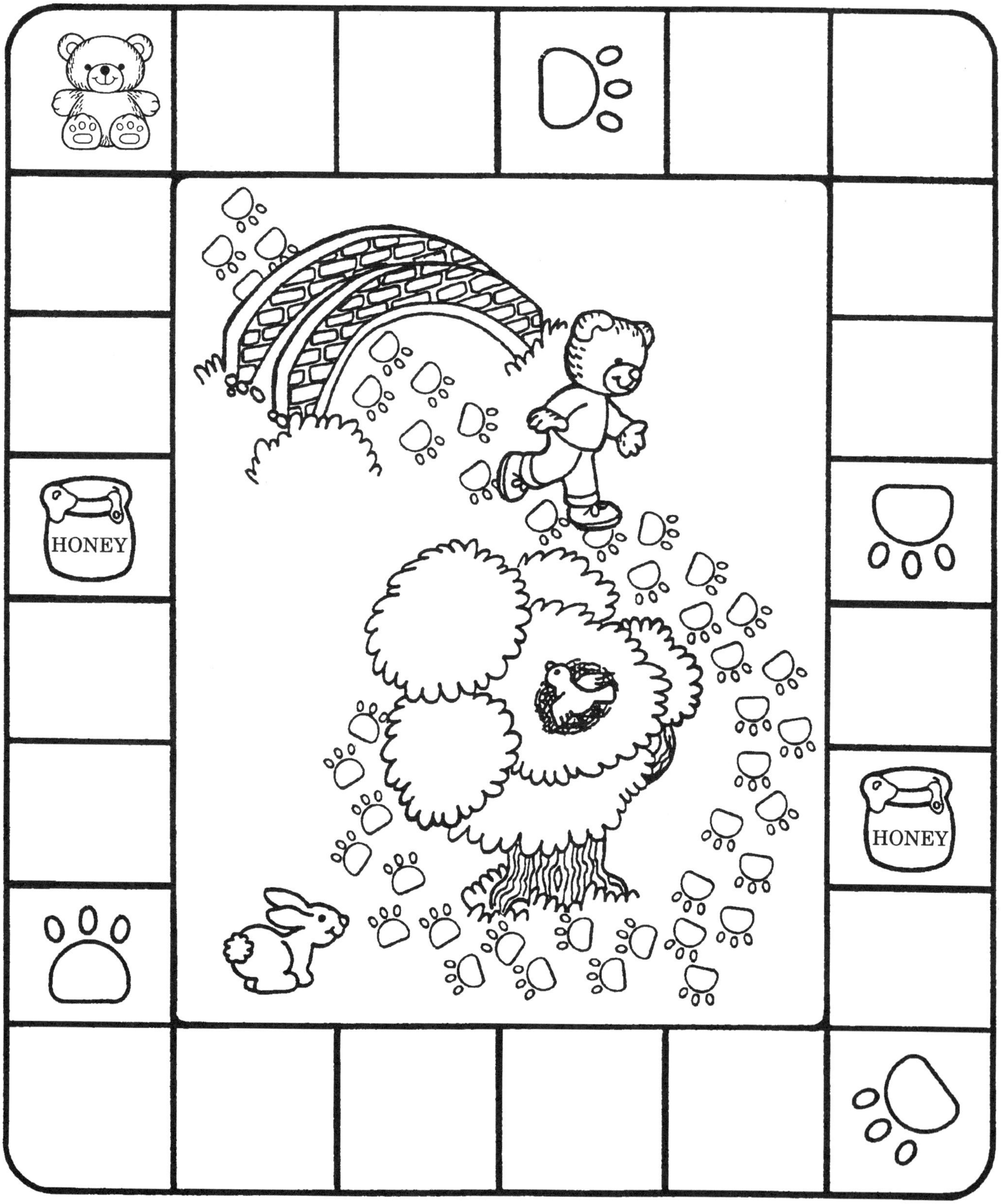

Teaching Aid 5: Cards for Bear Hunt: Deck 1

3	3	3	3	3
4	4	4	4	4
5	5	5	5	5
6	6	6	6	6
7	7	7	7	7
7	7	7	8	8
8	8	8	8	8
8	9	9	9	9

Teaching Aid 6: Cards for Bear Hunt: Deck 2

6	7	8	8	9
9	9	10	10	10
11	11	11	11	11
11	11	12	12	12
12	12	12	12	13
13	13	13	13	13
13	14	14	14	14
14	15	16	17	18

Teaching Aid 7: Hundred Chart

1	2	3	4	5	6	7	8	9	10
11	12	13	14	15	16	17	18	19	20
21	22	23	24	25	26	27	28	29	30
31	32	33	34	35	36	37	38	39	40
41	42	43	44	45	46	47	48	49	50
51	52	53	54	55	56	57	58	59	60
61	62	63	64	65	66	67	68	69	70
71	72	73	74	75	76	77	78	79	80
81	82	83	84	85	86	87	88	89	90
91	92	93	94	95	96	97	98	99	100

Peephole Cards

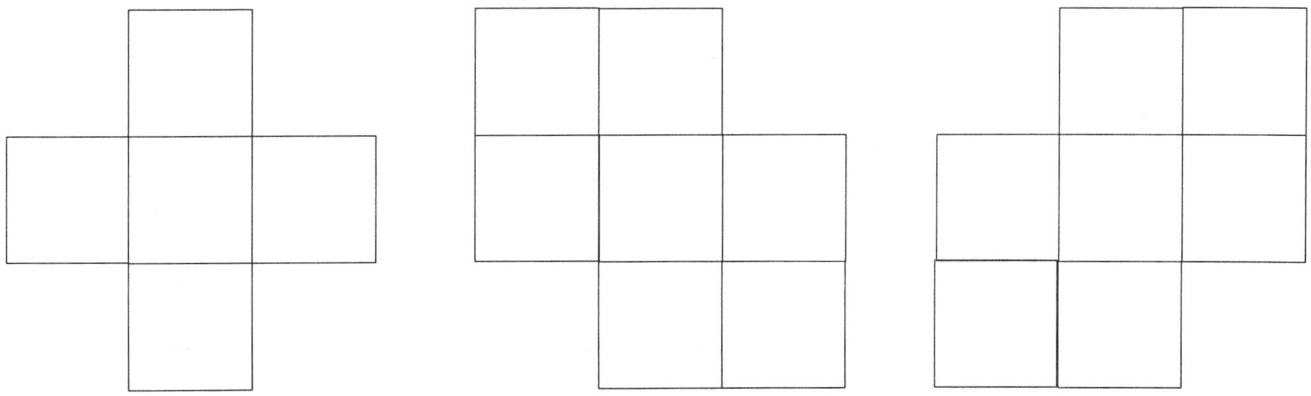

Teaching Aid 8: Graphing Paper

Teaching Aid 9: Water Rafts & Pup Tents

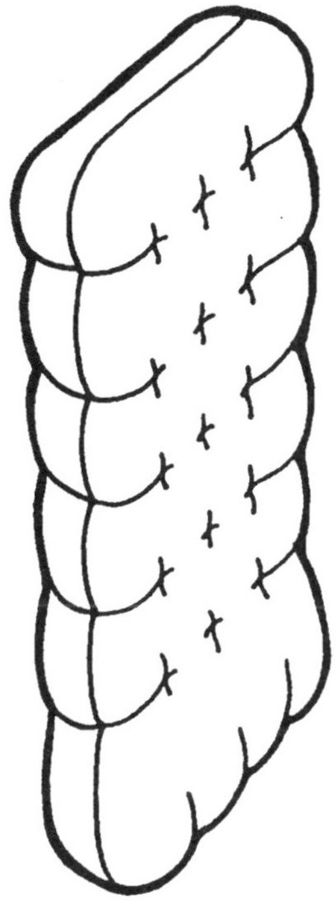
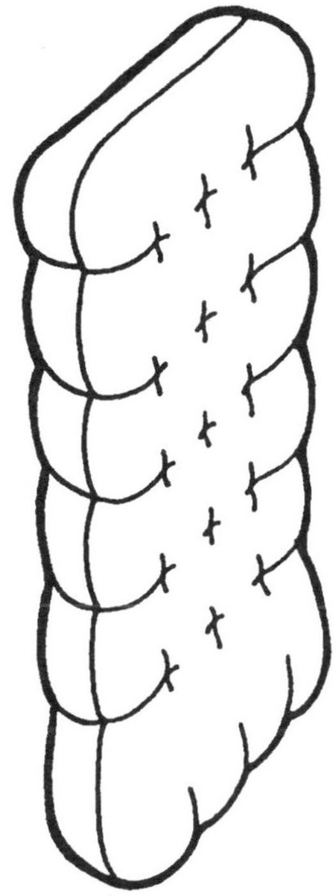
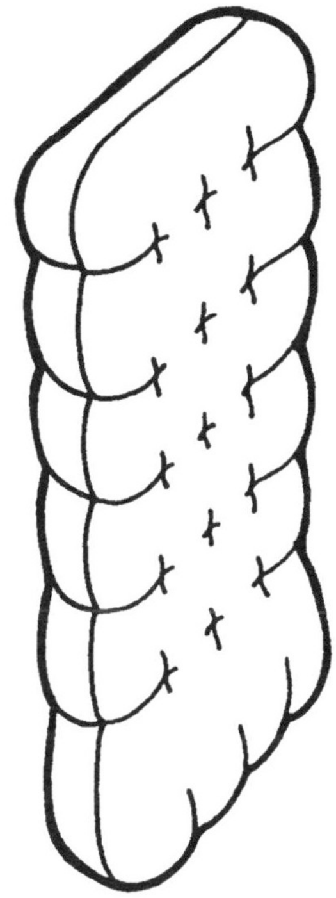

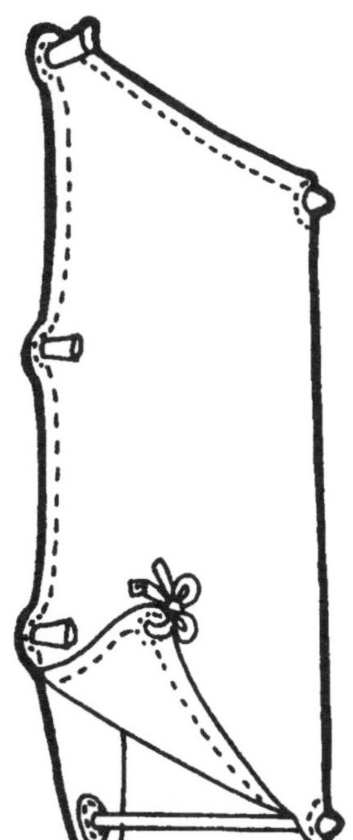
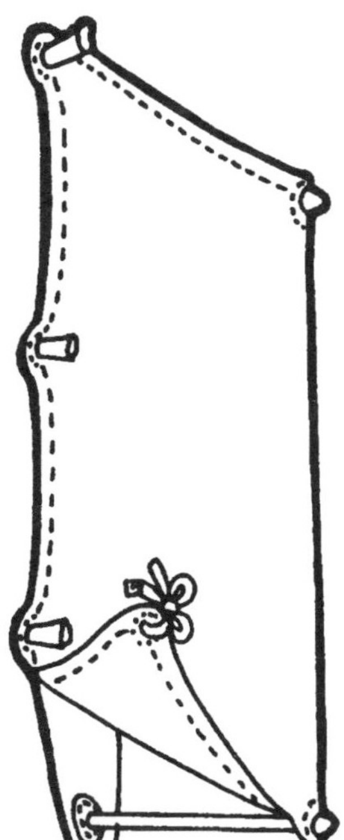
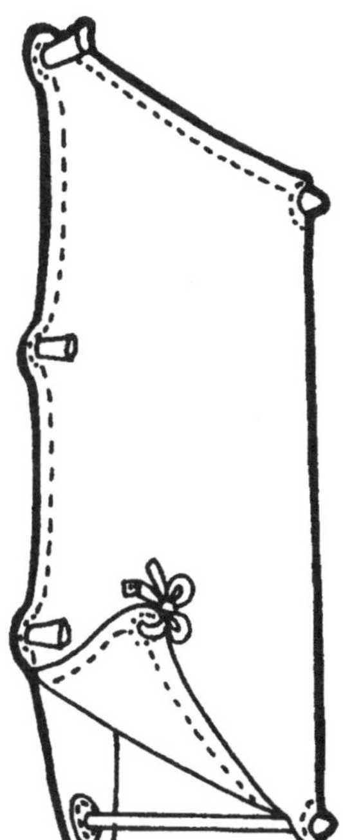